NATEF Standards Job Sheets

Engine Performance (A8)

Fourth Edition

Jack Erjavec
Ken Pickerill

Australia • Brazil • Japan • Korea • Mexico • Singapore • Spain • United Kingdom • United States

NATEF Standards Job Sheets
Engine Performance (A8)
Fourth Edition
Jack Erjavec & Ken Pickerill

VP, General Manager, Skills and Planning: Dawn Gerrain

Director, Development, Global Product Management, Skills: Marah Bellegarde

Product Manager: Erin Brennan

Senior Product Development Manager: Larry Main

Senior Content Developer: Meaghan Tomaso

Product Assistant: Maria Garguilo

Marketing Manager: Linda Kuper

Market Development Manager: Jonathon Sheehan

Senior Production Director: Wendy Troeger

Production Manager: Mark Bernard

Content Project Management: S4Carlisle

Art Direction: S4Carlisle

Media Developer: Debbie Bordeaux

Cover image(s): © Snaprender/Dreamstime.com
© Shutterstock.com/gameanna
© i3alda/www.fotosearch.com
© IStockPhoto.com/tarras79
© IStockPhoto.com/tarras79
© Laurie Entringer
© IStockPhoto.com/Zakai
© stoyanh/www.fotosearch.com

Library of Congress Control Number: 2014931168

ISBN-13: 978-1-1116-4704-9
ISBN-10: 1-111-64704-6

Cengage Learning
200 First Stamford Place, 4th Floor
Stamford, CT 06902
USA

Notice to the Reader
Publisher does not warrant or guarantee any of the products described herein or perform any independent analysis in connection with any of the product information contained herein. Publisher does not assume, and expressly disclaims, any obligation to obtain and include information other than that provided to it by the manufacturer. The reader is expressly warned to consider and adopt all safety precautions that might be indicated by the activities described herein and to avoid all potential hazards. By following the instructions contained herein, the reader willingly assumes all risks in connection with such instructions. The publisher makes no representations or warranties of any kind, including but not limited to, the warranties of fitness for particular purpose or merchantability, nor are any such representations implied with respect to the material set forth herein, and the publisher takes no responsibility with respect to such material. The publisher shall not be liable for any special, consequential, or exemplary damages resulting, in whole or part, from the readers' use of, or reliance upon, this material.

Printed in the United States of America
1 2 3 4 5 6 7 18 17 16 15 14

CONTENTS

NATEF TASK LIST FOR ENGINE PERFORMANCE

Required Supplemental Tasks (RST) List

Shop and Personal Safety

1. Identify general shop safety rules and procedures.
2. Utilize safe procedures for handling of tools and equipment.
3. Identify and use proper placement of floor jacks and jack stands.
4. Identify and use proper procedures for safe lift operation.
5. Utilize proper ventilation procedures for working within the lab/shop area.
6. Identify marked safety areas.
7. Identify the location and the types of fire extinguishers and other fire safety equipment; demonstrate knowledge of the procedures for using fire extinguishers and other fire safety equipment.
8. Identify the location and use of eyewash stations.
9. Identify the location of the posted evacuation routes.
10. Comply with the required use of safety glasses, ear protection, gloves, and shoes during lab/shop activities.
11. Identify and wear appropriate clothing for lab/shop activities.
12. Secure hair and jewelry for lab/shop activities.
13. Demonstrate awareness of the safety aspects of supplemental restraint systems (SRS), electronic brake control systems, and hybrid vehicle high-voltage circuits.
14. Demonstrate awareness of the safety aspects of high-voltage circuits (such as high intensity discharge (HID) lamps, ignition systems, injection systems, etc.).
15. Locate and demonstrate knowledge of material safety data sheets (MSDS).

Tools and Equipment

1. Identify tools and their usage in automotive applications.
2. Identify standard and metric designation.
3. Demonstrate safe handling and use of appropriate tools.
4. Demonstrate proper cleaning, storage, and maintenance of tools and equipment.
5. Demonstrate proper use of precision measuring tools (i.e., micrometer, dial-indicator, dial-caliper).

Preparing Vehicle for Service

1. Identify information needed and the service requested on a repair order.
2. Identify purpose and demonstrate proper use of fender covers, mats.
3. Demonstrate use of the three C's (concern, cause, and correction).
4. Review vehicle service history.
5. Complete work order to include customer information, vehicle identifying information, customer concern, related service history, cause, and correction.

Preparing Vehicle for Customer

1. Ensure vehicle is prepared to return to customer per school/company policy (floor mats, steering wheel cover, etc.).

Maintenance and Light Repair (MLR) Task List

A. General

A.1. Research applicable vehicle and service information, vehicle service history, service precautions, and technical service bulletins. — Priority Rating 1
A.2. Perform engine absolute (vacuum/boost) manifold pressure tests; determine necessary action. — Priority Rating 1
A.3. Perform cylinder power balance test; determine necessary action. — Priority Rating 2
A.4. Perform cylinder cranking and running compression tests; determine necessary action. — Priority Rating 1
A.5. Perform cylinder leakage test; determine necessary action. — Priority Rating 1
A.6. Verify engine operating temperature. — Priority Rating 1
A.7. Remove and replace spark plugs; inspect secondary ignition components for wear and damage. — Priority Rating 1

B. Computerized Controls

B.1. Retrieve and record diagnostic trouble codes, OBD monitor status, and freeze frame data; clear codes when applicable. — Priority Rating 1
B.2. Describe the importance of running all OBDII monitors for repair verification. — Priority Rating 1

C. Fuel, Air, Induction, and Exhaust Systems

C.1. Replace fuel filter(s). — Priority Rating 1
C.2. Inspect, service, or replace air filters, filter housings, and intake duct work. — Priority Rating 1
C.3. Inspect integrity of the exhaust manifold, exhaust pipes, muffler(s), catalytic converter(s), resonator(s), tail pipe(s), and heat shields; determine necessary action. — Priority Rating 1
C.4. Inspect condition of exhaust system hangers, brackets, clamps, and heat shields; repair or replace as needed. — Priority Rating 1
C.5. Check and refill diesel exhaust fluid (DEF). — Priority Rating 3

D. Emissions Control Systems

D.1. Inspect, test, and service positive crankcase ventilation (PCV) filter/breather cap, valve, tubes, orifices, and hoses; perform necessary action. — Priority Rating 2

Automobile Service Technology (AST) Task List

A. General Engine Diagnosis

A.1.	Identify and interpret engine performance concerns; determine necessary action.	Priority Rating 1
A.2.	Research applicable vehicle and service information, vehicle service history, service precautions, and technical service bulletins.	Priority Rating 1
A.3.	Diagnose abnormal engine noises or vibration concerns; determine necessary action.	Priority Rating 3
A.4.	Diagnose the cause of excessive oil consumption, coolant consumption, unusual exhaust color, odor, and sound; determine necessary action.	Priority Rating 2
A.5.	Perform engine absolute (vacuum/boost) manifold pressure tests; determine necessary action.	Priority Rating 1
A.6.	Perform cylinder power balance test; determine necessary action.	Priority Rating 2
A.7.	Perform cylinder cranking and running compression tests; determine necessary action.	Priority Rating 1
A.8.	Perform cylinder leakage test; determine necessary action.	Priority Rating 1
A.9.	Diagnose engine mechanical, electrical, electronic, fuel, and ignition concerns; determine necessary action.	Priority Rating 2
A.10.	Verify engine operating temperature; determine necessary action.	Priority Rating 1
A.11.	Verify correct camshaft timing.	Priority Rating 1

B. Computerized Engine Controls Diagnosis and Repair

B.1.	Retrieve and record diagnostic trouble codes, OBD monitor status, and freeze frame data; clear codes when applicable.	Priority Rating 1
B.2.	Access and use service information to perform step-by-step (troubleshooting) diagnosis.	Priority Rating 1
B.3.	Perform active tests of actuators using a scan tool; determine necessary action.	Priority Rating 2
B.4.	Describe the importance of running all OBDII monitors for repair verification.	Priority Rating 1

C. Ignition System Diagnosis and Repair

C.1.	Diagnose (troubleshoot) ignition system–related problems such as no-starting, hard starting, engine misfire, poor driveability, spark knock, power loss, poor mileage, and emissions concerns; determine necessary action.	Priority Rating 2
C.2.	Inspect and test crankshaft and camshaft position sensor(s); perform necessary action.	Priority Rating 1
C.3.	Inspect, test, and/or replace ignition control module, powertrain/engine control module; reprogram as necessary.	Priority Rating 3
C.4.	Remove and replace spark plugs; inspect secondary ignition components for wear and damage.	Priority Rating 1

D. Fuel, Air Induction, and Exhaust Systems Diagnosis and Repair

D.1.	Check fuel for contaminants; determine necessary action.	Priority Rating 2
D.2.	Inspect and test fuel pumps and pump control systems for pressure, regulation, and volume; perform necessary action.	Priority Rating 1
D.3.	Replace fuel filter(s).	Priority Rating 1
D.4.	Inspect, service, or replace air filters, filter housings, and intake duct work.	Priority Rating 1
D.5.	Inspect throttle body, air induction system, intake manifold and gaskets for vacuum leaks and/or unmetered air.	Priority Rating 2
D.6.	Inspect and test fuel injectors.	Priority Rating 2

D.7. Verify idle control operation. Priority Rating 1

D.8. Inspect integrity of the exhaust manifold, exhaust pipes, muffler(s), catalytic converter(s), resonator(s), tail pipe(s), and heat shields; perform necessary action. Priority Rating 1

D.9. Inspect condition of exhaust system hangers, brackets, clamps, and heat shields; repair or replace as needed. Priority Rating 1

D.10. Perform exhaust system back-pressure test; determine necessary action. Priority Rating 2

D.11. Check and refill Diesel Exhaust Fluid (DEF). Priority Rating 3

E. Emissions Control Systems Diagnosis and Repair

E.1. Diagnose oil leaks, emissions, and driveability concerns caused by the positive crankcase ventilation (PCV) system; determine necessary action. Priority Rating 3

E.2. Inspect, test, and service positive crankcase ventilation (PCV) filter/breather cap, valve, tubes, orifices, and hoses; perform necessary action. Priority Rating 2

E.3. Diagnose emissions and driveability concerns caused by the exhaust gas recirculation (EGR) system; determine necessary action. Priority Rating 3

E.4. Inspect, test, service, and replace components of the EGR system including tubing, exhaust passages, vacuum/pressure controls, filters, and hoses; perform necessary action. Priority Rating 2

E.5. Inspect and test electrical/electronically operated components and circuits of air injection systems; perform necessary action. Priority Rating 3

E.6. Inspect and test catalytic converter efficiency. Priority Rating 2

E.7. Inspect and test components and hoses of the evaporative emissions control system; perform necessary action. Priority Rating 1

E.8. Interpret diagnostic trouble codes (DTCs) and scan tool data related to the emissions control systems; determine necessary action. Priority Rating 3

Master Automobile Service Technology (MAST) Task List

A. General Engine Diagnosis

A.1. Identify and interpret engine performance concerns; determine necessary action. Priority Rating 1

A.2. Research applicable vehicle and service information, vehicle service history, service precautions, and technical service bulletins. Priority Rating 1

A.3. Diagnose abnormal engine noises or vibration concerns; determine necessary action. Priority Rating 3

A.4. Diagnose the cause of excessive oil consumption, coolant consumption, unusual exhaust color, odor, and sound; determine necessary action. Priority Rating 2

A.5. Perform engine absolute (vacuum/boost) manifold pressure tests; determine necessary action. Priority Rating 1

A.6. Perform cylinder power balance test; determine necessary action. Priority Rating 2

A.7. Perform cylinder cranking and running compression tests; determine necessary action. Priority Rating 1

A.8. Perform cylinder leakage test; determine necessary action. Priority Rating 1

A.9. Diagnose engine mechanical, electrical, electronic, fuel, and ignition concerns; determine necessary action. Priority Rating 2

A.10. Verify engine operating temperature; determine necessary action. Priority Rating 1

A.11. Verify correct camshaft timing. Priority Rating 1

B. Computerized Engine Controls Diagnosis and Repair

- B.1. Retrieve and record diagnostic trouble codes, OBD monitor status, and freeze frame data; clear codes when applicable. Priority Rating 1
- B.2. Access and use service information to perform step-by-step (troubleshooting) diagnosis. Priority Rating 1
- B.3. Perform active tests of actuators using a scan tool; determine necessary action. Priority Rating 2
- B.4. Describe the importance of running all OBDII monitors for repair verification. Priority Rating 1
- B.5. Diagnose the causes of emissions or driveability concerns with stored or active diagnostic trouble codes; obtain, graph, and interpret scan tool data. Priority Rating 1
- B.6. Diagnose emissions or driveability concerns without stored diagnostic trouble codes; determine necessary action. Priority Rating 1
- B.7. Inspect and test computerized engine control system sensors, powertrain/engine control module (PCM/ECM), actuators, and circuits using a graphing multimeter (GMM)/digital storage oscilloscope (DSO); perform necessary action. Priority Rating 2
- B.8. Diagnose driveability and emissions problems resulting from malfunctions of interrelated systems (cruise control, security alarms, suspension controls, traction controls, A/C, automatic transmissions, non-OEM installed accessories, or similar systems); determine necessary action. Priority Rating 3

C. Ignition System Diagnosis and Repair

- C.1. Diagnose (troubleshoot) ignition system–related problems such as no-starting, hard starting, engine misfire, poor driveability, spark knock, power loss, poor mileage, and emissions concerns; determine necessary action. Priority Rating 2
- C.2. Inspect and test crankshaft and camshaft position sensor(s); perform necessary action. Priority Rating 1
- C.3. Inspect, test, and/or replace ignition control module, powertrain/engine control module; reprogram as necessary. Priority Rating 3
- C.4. Remove and replace spark plugs; inspect secondary ignition components for wear and damage. Priority Rating 1

D. Fuel, Air Induction, and Exhaust Systems Diagnosis and Repair

- D.1. Diagnose (troubleshoot) hot or cold no-starting, hard starting, poor driveability, incorrect idle speed, poor idle, flooding, hesitation, surging, engine misfire, power loss, stalling, poor mileage, dieseling, and emissions problems; determine necessary action. Priority Rating 2
- D.2. Check fuel for contaminants; determine necessary action. Priority Rating 2
- D.3. Inspect and test fuel pumps and pump control systems for pressure, regulation, and volume; perform necessary action. Priority Rating 1
- D.4. Replace fuel filter(s). Priority Rating 1
- D.5. Inspect, service, or replace air filters, filter housings, and intake duct work. Priority Rating 1
- D.6. Inspect throttle body, air induction system, intake manifold and gaskets for vacuum leaks and/or unmetered air. Priority Rating 2
- D.7. Inspect and test fuel injectors. Priority Rating 2
- D.8. Verify idle control operation. Priority Rating 1

D.9. Inspect integrity of the exhaust manifold, exhaust pipes, muffler(s), catalytic converter(s), resonator(s), tail pipe(s), and heat shields; perform necessary action. Priority Rating 1
D.10. Inspect condition of exhaust system hangers, brackets, clamps, and heat shields; repair or replace as needed. Priority Rating 1
D.11. Perform exhaust system back-pressure test; determine necessary action. Priority Rating 2
D.12. Check and refill diesel exhaust fluid (DEF). Priority Rating 3
D.13. Test the operation of turbocharger/supercharger systems; determine necessary action. Priority Rating 3

E. Emissions Control Systems Diagnosis and Repair

E.1. Diagnose oil leaks, emissions, and driveability concerns caused by the positive crankcase ventilation (PCV) system; determine necessary action. Priority Rating 3
E.2. Inspect, test, and service positive crankcase ventilation (PCV) filter/breather cap, valve, tubes, orifices, and hoses; perform necessary action. Priority Rating 2
E.3. Diagnose emissions and driveability concerns caused by the exhaust gas recirculation (EGR) system; determine necessary action. Priority Rating 3
E.4. Diagnose emissions and driveability concerns caused by the secondary air injection and catalytic converter systems; determine necessary action. Priority Rating 2
E.5. Diagnose emissions and driveability concerns caused by the evaporative emissions control system; determine necessary action. Priority Rating 2
E.6. Inspect and test electrical/electronic sensors, controls, and wiring of exhaust gas recirculation (EGR) systems; perform necessary action. Priority Rating 2
E.7. Inspect, test, service, and replace components of the EGR system including tubing, exhaust passages, vacuum/pressure controls, filters, and hoses; perform necessary action. Priority Rating 2
E.8. Inspect and test electrical/electronically operated components and circuits of air injection systems; perform necessary action. Priority Rating 3
E.9. Inspect and test catalytic converter efficiency. Priority Rating 2
E.10. Inspect and test components and hoses of the evaporative emissions control system; perform necessary action. Priority Rating 1
E.11. Interpret diagnostic trouble codes (DTCs) and scan tool data related to the emissions control systems; determine necessary action. Priority Rating 3

DEFINITION OF TERMS USED IN THE TASK LIST

To clarify the intent of these tasks, NATEF has defined some of the terms used in the task listings. To get a good understanding of what the task includes, refer to this glossary while reading the task list.

adjust To bring components to specified operational settings.
analyze Assess the condition of a component or system.
assemble (reassemble) To fit together the components of a device or system.
bleed To remove air from a closed system.
CAN Acronym for *Controller Area Network*. CAN is a network protocol (SAE J2284/ISO 15765-4) used to interconnect a network of electronic control modules.
charge To bring to a specified state (e.g., battery or air conditioning system).
check To verify condition by performing an operational or comparative examination.

clean To rid components of foreign matter for the purpose of reconditioning, repairing, measuring, and reassembling.

confirm To acknowledge something has happened with firm assurance.

demonstrate To show or exhibit the knowledge of a theory or procedure.

determine To establish the procedure to be used to perform the necessary repair.

determine necessary action Indicates that the diagnostic routine(s) is the primary emphasis of a task. The student is required to perform the diagnostic steps and communicate the diagnostic outcomes and corrective actions required, addressing the concern or problem. The training program determines the communication method (worksheet, test, verbal communication, or other means deemed appropriate) and whether the corrective procedures for these tasks are actually performed.

diagnose To identify the cause of a problem.

differentiate To perceive the difference in or between one thing to other things.

disassemble To separate a component's parts as a preparation for cleaning, inspection, or service.

discharge To empty a storage device or system.

high voltage Voltages of 50 volts or higher.

identify To establish the identity of a vehicle or component prior to service; to determine the nature or degree of a problem.

inspect (see *check*)

install (reinstall) To place a component in its proper position in a system.

jump-start To use an auxiliary power supply to assist a battery to crank an engine.

listen To use audible clues in the diagnostic process; to hear the customer's description of a problem.

locate Determine or establish a specific spot or area.

maintain To keep something at a specified level, position, rate, etc.

network A system of interconnected electrical modules or devices.

on-board diagnostics (OBD) Diagnostic protocol that monitors computer inputs and outputs for failures.

parasitic draw Electrical loads that are still present when the ignition circuit is OFF.

perform To accomplish a procedure in accordance with established methods and standards.

perform necessary action Indicates that the student is to perform the diagnostic routine(s) and perform the corrective action item. Where various scenarios (conditions or situations) are presented in a single task, at least one of the scenarios must be accomplished.

pressure test To use air or fluid pressure to determine the integrity, condition, or operation of a component or system.

priority ratings Indicates the minimum percentage of tasks, by area, that a program must include in its curriculum in order to be certified in that area.

reassemble (see *assemble*)

remove To disconnect and separate a component from a system.

repair To restore a malfunctioning component or system to operating condition.

replace	To exchange a component; to reinstall a component.
research	To make a thorough investigation into a situation or matter.
service	To perform a procedure as specified in the owner's or service manual.
test	To verify a condition through the use of meters, gauges, or instruments.
torque	To tighten a fastener to a specified degree or tightness (in a given order or pattern if multiple fasteners are involved on a single component).
verify	To confirm that a problem exists after hearing the customer's concern; or, to confirm the effectiveness of a repair.
voltage drop	A reduction in voltage (electrical pressure) caused by the resistance in a component or circuit.

CROSS-REFERENCE GUIDES

RST Task	**Job Sheet**
Shop and Personal Safety	
1.	1
2.	2
3.	2
4.	2
5.	2
6.	2
7.	3
8.	1
9.	2
10.	1
11.	1
12.	1
13.	4, 5
14.	6
15.	7
Tools and Equipment	
1.	8
2.	8
3.	8
4.	8
5.	8
Preparing Vehicle for Service	
1.	9
2.	9
3.	9
4.	9
5.	9
Preparing Vehicle for Customer	
1.	9

MLR Task	**Job Sheet**
A.1	11
A.2	12
A.3	13
A.4	14

MLR Task	Job Sheet
A.5	15
A.6	17
A.7	30
B.1	19
B.2	22
C.1	34
C.2	35
C.3	38
C.4	38
C.5	40
D.1	32

AST Task	Job Sheet
A.1	10
A.2	11
A.3	10
A.4	10
A.5	12
A.6	13
A.7	14
A.8	15
A.9	16
A.10	17
A.11	18
B.1	19
B.2	20
B.3	21
B.4	22
C.1	27
C.2	28
C.3	29
C.4	30
D.1	32
D.2	33
D.3	34
D.4	35
D.5	35
D.6	36
D.7	37
D.8	38
D.9	38
D.10	39
D.11	40
E.1	42
E.2	42
E.3	43
E.4	43
E.5	44
E.6	39
E.7	45
E.8	45

MAST Task	Job Sheet
A.1	10
A.2	11
A.3	10
A.4	10
A.5	12
A.6	13
A.7	14
A.8	15
A.9	16
A.10	17
A.11	18
B.1	19
B.2	20
B.3	21
B.4	22
B.5	23
B.6	24
B.7	25
B.8	26
C.1	27
C.2	28
C.3	29
C.4	30
D.1	31
D.2	32
D.3	33
D.4	34
D.5	35
D.6	35
D.7	36
D.8	37
D.9	38
D.10	38
D.11	39
D.12	40
D.13	41
E.1	42
E.2	42
E.3	43
E.4	44
E.5	45
E.6	43
E.7	43
E.8	44
E.9	39
E.10	45
E.11	45

PREFACE

The automotive service industry continues to change with the technological changes made by automobile, tool, and equipment manufacturers. Today's automotive technician must have a thorough knowledge of automotive systems and components, good computer skills, exceptional communication skills, good reasoning, the ability to read and follow instructions, and above average mechanical aptitude and manual dexterity.

This new edition, like the last, was designed to give students a chance to develop the same skills and gain the same knowledge that today's successful technician has. This edition also reflects the changes in the guidelines established by the National Automotive Technicians Education Foundation (NATEF) in 2013.

The purpose of NATEF is to evaluate technician training programs against standards developed by the automotive industry and recommend qualifying programs for certification (accreditation) by ASE (National Institute for Automotive Service Excellence). Programs can earn ASE certification upon the recommendation of NATEF. NATEF's national standards reflect the skills that students must master. ASE certification through NATEF evaluation ensures that certified training programs meet or exceed industry-recognized, uniform standards of excellence.

At the expense of much time and thought, NATEF has assembled a list of basic tasks for each of their certification areas. These tasks identify the basic skills and knowledge levels that competent technicians have. The tasks also identify what is required for a student to start a successful career as a technician.

In June 2013, after many discussions with the industry, NATEF established a new model for automobile program standards. This new model is reflected in this edition and covers the new standards that are based on three (3) levels: Maintenance & Light Repair (MLR), Automobile Service Technician (AST), and Master Automobile Service Technician (MAST). Each successive level includes all the tasks of the previous level in addition to new tasks. In other words, the AST task list includes all of the MLR tasks plus additional tasks. The MAST task list includes all of AST tasks plus additional tasks specifically for MAST.

Most of the content in this book are job sheets. These job sheets relate to the tasks specified by NATEF, according to the appropriate certification level. The main considerations during the creation of these job sheets were student learning and program certification by NATEF. Students are guided through standard industry-accepted procedures. While they are progressing, they are asked to report their findings as well as offer their thoughts on the steps they have just completed. The questions asked of the students are thought provoking and require students to apply what they know to what they observe.

The job sheets were also designed to be generic. That is, whenever possible, the tasks can be performed on any vehicle from any manufacturer. Also, completion of the sheets does not require the use of specific brands of tools and equipment. Rather, students use what is available. In addition, the job sheets can be used as a supplement to any good textbook.

Also included are description and basic use of the tools and equipment listed in NATEF's standards. The standards recognize that not all programs have the same needs, nor do all programs teach all of the NATEF tasks. Therefore, the basic philosophy for the tools and equipment requirement is that the training should be as thorough as possible with the tools and equipment necessary for those tasks.

Theory instruction and hands-on experience of the basic tasks provide initial training for employment in automotive service or further training in any or all of the specialty areas. Competency in the tasks indicates to employers that you are skilled in that area. You need to know the appropriate theory, safety, and support information for each required task. This should include identification and use of the required tools, testing, and measurement equipment required for the tasks, the use of current reference and training materials, the proper way to write work orders and warranty reports, and the storage, handling, and use of hazardous materials as required by the "Right to Know" law, and federal, state, and local governments.

Words to the Instructor: We suggest you grade these job sheets based on completion and reasoning. Make sure the students answer all questions. Then, look at their reasoning to see if the task was actually completed and to get a feel for their understanding of the topic. It will be easy for students to copy others' measurements and findings, but each student should have their own base of understanding and that will be reflected in their explanations.

Words to the Student: While completing the job sheets, you have a chance to develop the skills you need to be successful. When asked for your thoughts or opinions, think about what you observed. Think about what could have caused those results or conditions. You are not being asked to give accurate explanations for everything you do or observe. You are only asked to think. Thinking leads to understanding. Good technicians are good because they have a basic understanding of what they are doing and why they are doing it.

Jack Erjavec and Ken Pickerill

ENGINE PERFORMANCE SYSTEMS

To prepare you to learn what you should learn from completing the job sheets, some basics must be covered. This discussion begins with an overview of engine performance systems. Emphasis is placed on what they do and how they work. This includes the major components and designs of engine systems and their role in the efficient operation of engines of all designs.

Preparing to work on a vehicle would not be complete if certain safety issues were not addressed. This discussion covers those things you should and should not do while working on engine performance systems. Included are proper ways to deal with hazardous and toxic materials.

NATEF's task list for Engine Performance certification is also given with definitions of some of the terms used to describe the tasks. This list gives you a good look at what the experts say you need to know before you can be considered competent to work on engine performance systems.

Following the task list are descriptions of the various tools and types of equipment you need to be familiar with. These are the tools you will use to complete the job sheets. They are also the tools NATEF has identified as being necessary for servicing engine performance systems.

Figure 1 A spark plug.

Following the tool discussion is a cross-reference guide that shows what NATEF tasks are related to specific job sheets. In most cases there are single job sheets for each task. Some tasks are part of a procedure and when that occurs, one job sheet may cover two or more tasks. The remainder of the book contains the job sheets.

BASIC ENGINE PERFORMANCE THEORY

In order for an engine to run efficiently, complete combustion must take place in each of its cylinders. To have complete combustion, the cylinders must receive the correct amount of fuel mixed with the correct amount of air. Then, that mixture must be shocked with the right amount of heat at the correct time. An engine is fitted with many and various systems that work toward achieving the goal of total combustion.

Ignition Systems

One of the requirements for an efficiently running engine is the correct amount of heat delivered into the cylinders at the right time. This requirement is the responsibility of the ignition system. The ignition system supplies properly timed, high-voltage surges to the spark plugs (Figure 1). These voltage surges cause combustion inside the cylinder.

For each cylinder in an engine, the ignition system has three main jobs. First, it must generate an electrical spark that has enough heat to ignite the air/fuel mixture in the combustion chamber. Second, it must maintain that spark long enough to allow for the combustion of all the air and fuel in the cylinders. Finally, it must deliver the spark to each cylinder so that combustion can begin at the right time during the compression stroke of each cylinder.

Ignition Timing

Ignition timing refers to the precise time that spark occurs. If optimum engine performance is to be maintained, the ignition timing of the engine must change as the operating conditions of the engine change.

At higher engine speeds, the crankshaft turns through more degrees in a given period of time. If combustion is to be completed at the correct time, ignition timing must occur sooner or be advanced.

However, air/fuel mixture turbulence increases with rpm. This causes the mixture, inside the cylinder, to burn faster. Increased turbulence requires that ignition must occur slightly later or be slightly retarded. During heavy loads, the mixture burns faster so the ignition timing must again be retarded or combustion will end too soon. Achieving correct ignition timing is the result of careful balancing of engine operating conditions, especially speed and load.

Firing Order

Each cylinder of an engine produces power once every 720 degrees of crankshaft rotation. Each cylinder must have a power stroke at its own appropriate time during the rotation. To make this possible, the pistons and rods are arranged in a precise fashion.

The ignition system must be able to monitor the rotation of the crankshaft and the relative position of each piston in order to determine which piston is on its compression stroke. It must also be able to deliver a high-voltage surge to each cylinder at the proper time during its compression stroke.

Basic Ignition System

All ignition systems consist of two interconnected electrical circuits: a primary (low voltage) circuit and a secondary (high voltage) circuit. Depending on the exact type of ignition system, components in the primary circuit include the battery, ignition switch, ignition coil primary winding, triggering device, and a switching device or control module. The secondary circuit includes the ignition coil secondary winding, distributor cap and rotor (older systems), ignition (spark plug) cables, and spark plugs.

Primary Circuit Operation

When the ignition switch is in the on position, current from the battery flows through the ignition switch to the primary winding of the ignition coil. From there it passes through some type of switching device and back to ground. The current flow in the ignition coil's primary winding creates a magnetic field. The switching device or control module interrupts this current flow at predetermined times. When it does, the magnetic field in the primary winding collapses. This collapse generates a high-voltage surge in the secondary winding of the ignition coil.

Secondary Circuit Operation

The secondary circuit carries high voltage to the spark plugs. The exact manner in which the secondary circuit delivers these high-voltage surges depends on the system design. Until 1984 all ignition systems used some type of distributor to accomplish this job. However, in an effort to reduce emissions, improve fuel economy, and boost component reliability, auto manufacturers are now using distributor-less or Electronic Ignition (EI) systems.

In a Distributor Ignition (DI) system, high voltage from the secondary winding passes through an ignition cable running from the coil to the distributor. The distributor then distributes the high voltage to the individual spark plugs through a set of ignition cables. The cables are arranged in the distributor cap according to the firing order of the engine. A rotor, driven by the distributor shaft, rotates and completes the electrical path from the secondary winding of the coil to the individual spark plugs.

EI systems have no distributor; instead, spark distribution is controlled by an electronic control unit and/or the vehicle's computer. Instead of a single ignition coil for all cylinders, each cylinder may have its own ignition coil, or two cylinders may share one coil. The coils are wired directly to the spark plug they control. An ignition control module controls the firing order and the spark timing and advance. In EI systems, a crank sensor located at the front of the crankshaft is used to trigger the ignition system. Most engines also have a camshaft sensor that is used to synchronize the firing of the spark plugs with the exact position of the pistons.

Ignition Coil Action

To generate a spark to begin combustion, the ignition system must deliver high voltage to the spark plugs. Because the amount of voltage required to bridge the gap of the spark plug varies with the operating conditions, ignition systems can easily supply 30,000 to 60,000 or more volts to force a spark across the air gap. Since the battery delivers 12 volts, a method of stepping up the voltage must be used. Multiplying battery voltage is the job of a coil.

The ignition coil is a pulse transformer. It transforms battery voltage into short bursts of high voltage.

In the center of the coil is a core made of laminated iron sheets. These sheets are shaped like the letter "E" and the primary and secondary windings are wound around the center of the E-core. The primary winding has approximately 200 turns of heavy wire and the two ends of this winding are connected to the primary terminals of the coil. These terminals are usually identified with positive and negative symbols. Enamel-type insulation prevents the primary windings from touching each other.

Secondary coil windings are made of very fine wire. These windings are on the inside of the primary winding. The ends of the secondary winding are usually connected to one of the primary terminals and to the high-tension terminal in the coil tower. Paper insulation is placed between the layers of windings. Current coils are air-cooled, however some older ones were oil-cooled.

Spark Plugs

Every type of ignition system uses spark plugs. The spark plugs provide the crucial air gap across which the high-voltage current from the coil flows across in the form of an arc. The three main parts of a spark plug are the steel core, the ceramic core or insulator, which acts as a heat conductor, and a pair of electrodes, one insulated in the core and the other grounded on the shell. The shell holds the ceramic core and electrodes in a gas-tight assembly and has the threads needed for plug installation in the engine. An ignition cable connects the secondary to the top of the plug. Current flows through the center of the plug and arcs from the tip of the center electrode to the ground electrode. The resulting spark ignites the air/fuel mixture in the combustion chamber. Most automotive spark plugs also have a resistor between the top terminal and the center electrode. This resistor reduces radio frequency interference (RFI), which prevents noise on stereo equipment. The resistor, like all other resistances in the secondary, increases the voltage needed to jump the gap of the spark plug.

Spark plugs come in many different sizes and designs to accommodate different engines. To fit properly, spark plugs must be of the proper size and reach. Another design factor that determines the usefulness of a spark plug for a specific application is its heat range. The desired heat range depends on the design of the engine and on the type of driving conditions the vehicle is subject to.

A terminal post on top of the center electrode is the point of contact for the spark plug cable. A ceramic insulator surrounds the center electrode, commonly made of a copper alloy, and a copper and glass seal is located between the electrode and the insulator. These seals prevent combustion gases from leaking out of the cylinder. Ribs on the insulator increase the distance between the terminal and the shell to help prevent electric arcing on the outside of the insulator. The steel spark plug shell is crimped over the insulation and a ground electrode, on the lower end of the shell, is positioned directly below the center electrode. There is an air gap between these two electrodes, and the width of this air gap is specified by the auto manufacturer.

Ignition Cables

Spark plug, or ignition, cables make up the secondary wiring. These cables carry the high voltage from the distributor or the multiple coils to the spark plugs. The cables are not solid wire; instead, they contain fiber cores that act as a resistor in the secondary circuit. They cut down on radio and television interference, increase firing voltages, and reduce spark plug wear by decreasing current. Metal terminals on each end of the spark plug wires contact the spark plug and the secondary terminal of the ignition coils or the distributor cap terminals. Insulated boots on the ends of the cables strengthen the connections as well as prevent dust and water infiltration and voltage loss. Coil-over-plug designs have the coil mount directly to the spark plug using only a boot between them, since these designs have a coil for each cylinder.

Spark Timing Systems

Electronic switching components can be located inside a separate housing known as an electronic control module or be part of the PCM. Today, timing changes are controlled by the PCM based on inputs from various sensors.

Based on the inputs it receives, the computer makes decisions regarding spark timing and sends signals to the ignition module to fire the spark plugs according to those inputs and according to the programs in its memory. Coil-over-plug systems often have the module incorporated into the ECM. In some coil-over-plug designs, the module is built into the coil itself, and may be able to report back to the PCM on the quality of the spark itself.

Crankshaft Position Sensors

The time when the primary circuit must be opened and closed is related to the position of the pistons and the crankshaft. Therefore, the position of the crankshaft is used to control the action of the switching unit.

A number of different types of sensors are used to monitor the position of the crankshaft. These crankshaft position sensors serve as triggering devices and include magnetic pulse generators, metal detection sensors, Hall-effect sensors, and photoelectric sensors.

The mounting location of these sensors depends on the design of the ignition system. All four types of sensors can be mounted in the distributor, which is turned by the camshaft.

Magnetic pulse generators and Hall-effect sensors can also be located on the crankshaft. These sensors are also commonly used on EI ignition systems. Both Hall-effect sensors and magnetic pulse generators can also be used as camshaft reference sensors to identify which cylinder is the next one to fire.

The magnetic pulse or PM generator operates on basic electromagnetic principles. Remember that a voltage is induced when a conductor moves through a magnetic field. The magnetic field is provided by the pick-up unit and the rotating timing disc provides the movement through the magnetic field needed to induce voltage.

The Hall-effect sensor produces a square wave signal that is more compatible with the digital signals required by on-board computers. The operation of a Hall-effect sensor is based on the Hall-effect principle, which states: If a current is allowed to flow through a thin conducting material, and that material is exposed to a magnetic field, voltage is produced.

The heart of the Hall generator is a thin semiconductor layer. Attached to it are two terminals—one positive, and the other negative—that are used to provide the source current for the Hall transformation.

Directly across from this semiconductor element is a permanent magnet. It is positioned so that its lines of flux bisect the Hall layer at right angles to the direction of current flow. Two additional terminals, located on either side of the Hall layer, form the signal output circuit.

When a moving metallic shutter blocks the magnetic field from reaching the Hall layer or element, the Hall-effect switch produces a voltage signal. When the shutter blade moves and allows the magnetic field to expand and reach the Hall element, the Hall-effect switch does not generate a voltage signal.

After leaving the Hall layer, the signal is routed to an amplifier where it is strengthened and inverted so that the signal reads high when it is actually coming in low and vice versa. Once it has been inverted, the signal goes through a Schmitt trigger, where it is turned into a clean square wave signal. After conditioning, the signal is sent to the switching unit.

Computer Controlled Ignition System Operation

Computer-controlled ignition systems control the primary circuit and distribute the firing voltages to the spark plugs.

Spark timing varies continuously to obtain optimum air/fuel combustion. The computer determines the best spark timing based on certain engine operating conditions, such as crankshaft position, engine speed, throttle position, engine coolant temperature, and initial and operating manifold or barometric pressure. Once the computer receives input from these and other sensors, it compares the existing operating conditions to information permanently stored or programmed into its memory. The computer matches the existing conditions to a set of conditions stored in its memory, determines proper timing setting, and sends a signal to the ignition module to fire the plugs.

Fuel Injection

In order for an engine to be efficient, it must receive the correct amount of fuel mixed with the correct amount of air. Providing this is the purpose of the fuel injection and fuel delivery systems.

In electronic fuel injection systems, the engine's fuel needs are measured by intake airflow past a sensor or by intake manifold pressure (vacuum). The airflow or manifold vacuum sensor converts its reading to an electrical signal and sends it to the PCM. Here the signal is processed and the current fuel needs are calculated. The PCM then sends an electrical signal to the fuel injectors. This signal determines the amount of time the injector opens and sprays fuel. This interval is known as the injector pulse width.

The PCM is self-regulating and controls the injectors on the basis of operating performance or parameters rather than on preprogrammed instructions. A PCM with a feedback loop, for example, reads the signals from the oxygen sensor, varies the pulse width of the injectors, and again reads the signals from the oxygen sensor. This is repeated until the injectors are pulsed for just the amount of time needed to get the proper amount of oxygen into the exhaust stream. While this interaction is occurring, the system is operating in closed loop. When conditions, such as starting or wide open throttle, demand that the signals from the oxygen sensor be ignored, the system operates in open loop. During open loop, injector pulse length is controlled by set parameters contained in the PCM's memory.

Many other sensors are used in fuel injection systems. Some of the more commonly used are an oxygen sensor, engine coolant temperature sensor, intake air temperature sensor, throttle position sensor, manifold absolute pressure sensor, mass airflow sensor, knock sensor, exhaust gas recirculation valve position sensor, and vehicle speed sensor.

Types of Fuel Injection

Many of the early EFI systems were Throttle Body Injection (TBI) systems in which the fuel was injected above the throttle plates. In these systems the injector assembly is located in the lower half of the intake manifold. Engines equipped with TBI were gradually equipped with Port Fuel Injection (PFI), which has injectors located in the intake ports of the cylinders. The latest gasoline injection type is direct fuel injection, or DFI. DFI injects fuel directly into the combustion chamber.

Throttle body injection systems have a throttle body assembly mounted on the intake manifold in the position previously occupied by a carburetor. The throttle body assembly usually contains one or two injectors.

On port fuel injection systems, fuel injectors are mounted at the back of each intake valve. Aside from the differences in injector location and number of injectors, operation of throttle body and port systems is quite similar with regard to fuel and air metering, sensors, and computer operation.

Throttle Body Fuel Injection

The basic TBI assembly consists of two major castings: a throttle body with a valve to control airflow and a fuel body to supply the required fuel. A fuel pressure regulator and fuel injector are integral parts of the fuel body. Also included as part of the assembly is a device to control idle speed and one to provide throttle valve positioning data.

The fuel pressure regulator used on the throttle body assembly is similar to a diaphragm-operated relief valve. Fuel pressure is on one side of the diaphragm and atmospheric pressure is on the other side. The regulator is designed to provide a constant pressure on the fuel injector throughout the range of engine loads and speeds. If regulator pressure is too high, a strong fuel odor is emitted and the engine runs too rich. On the other hand, regulator pressure that is too low results in poor engine performance or detonation can take place, due to the lean mixture.

The fuel injector is solenoid operated and pulsed on and off by the vehicle's engine control computer. When the injector's solenoid is energized, a normally closed ball valve is lifted and fuel under pressure is then injected at the walls of the throttle body bore just above the throttle plate.

A fuel injector has a movable armature in its center, and a pintle with a tapered tip is positioned at the lower end of the armature. A spring pushes the armature and pintle downward so the pintle tip seats in the discharge orifice. The injector coil surrounds the armature, and the two ends of the winding are connected to the terminals on the side of the injector. When the ignition switch is turned on, voltage is supplied to one injector terminal and the other terminal is connected through the computer. Each time the control unit completes the circuit from the injector winding to ground, current flows through the injector coil, and the coil magnetism moves the plunger and pintle upward. Under this condition, the pintle tip is unseated from the injector orifice, and fuel sprays out of this orifice.

Port Fuel Injection

PFI systems use one injector at each cylinder. They are mounted in the intake manifold near the cylinder head where they can inject a fine, atomized fuel mist as close as possible to the intake valve. Fuel lines run to each cylinder from a fuel manifold, usually referred to as a fuel rail. Since each cylinder has its own injector, fuel distribution is exactly equal.

The throttle body in a port fuel injection system controls the amount of air that enters the engine.

Port injectors have their tips located in the manifold where constant changes in vacuum would affect the amount of fuel injected. To compensate for these fluctuations, port injection systems are equipped with fuel pressure regulators that continually adjust the fuel pressure to maintain a constant pressure drop across the injector tips at all times. There are two types of fuel systems in use; the older return-line style fuel system and the latest type is the return-less fuel system.

On the return line style system, when fuel pressure reaches the setting of the regulator, a diaphragm moves against spring tension, and the regulator valve opens. This action allows fuel to flow through the return line to the fuel tank. The fuel pressure drops slightly when the pressure regulator valve opens, and in response, the spring closes the regulator valve.

A vacuum hose is connected from the intake manifold to the vacuum inlet on the pressure regulator. This hose supplies vacuum to the area where the diaphragm spring is located. The vacuum works with the fuel pressure to move the diaphragm and open the valve. When the engine is running at idle speed, high manifold vacuum is supplied to the regulator. Under this condition, the specified fuel pressure opens the regulator valve. When the engine is running under heavy load and/or wide-open throttle, a very low vacuum is supplied to the regulator. During these times, the vacuum does not help open the regulator valve and a higher fuel pressure is required to open the valve.

The change in fuel pressure allows the fuel to be sprayed into the manifold with the same effect, regardless of the pressure present in the manifold. When there is a high vacuum in the manifold, a very low pressure exists and the pressure difference between the fuel spray and the vacuum is the same as when there is a higher pressure in the manifold (low vacuum) and a higher fuel pressure.

Sequential fuel injection systems control each injector individually and according to the engine's firing order. Each injector is opened just before the intake valve opens, which means the mixture is never static in the intake manifold and adjustments to the mixture can be made almost instantaneously between the firing of one injector and the next.

Direct Fuel Injection

Similar to a diesel injection system, the DFI system delivers the fuel charge directly into the combustion chamber. The fuel pressure is much higher than the typical fuel injection systems and can reach 2,200 psi. Under this amount of pressure, the fuel is vaporized as it is injected into the cylinder. With DFI, mixtures can be substantially leaner (up to 35:1). This can improve fuel economy by as much as 30 percent and dramatically reduces hydrocarbon emissions on a cold engine. The introduction of the DFI systems are phasing out traditional port fuel systems and becoming commonplace as system application expands to more engine systems.

The DFI engine uses a special piston design, high-pressure swirl injectors, higher (often 12:1) compression ratios, and a special intake manifold; all designed to help burn a leaner fuel mixture than possible with conventional fuel injection. Since the injection of fuel is not dependent on the opening of the intake valve, the fuel can be injected during the compression stroke or the intake stroke, depending on the mode of operation. Since the intake carries only air, the intake can be enhanced for optimal performance.

The fuel pressure on the DFI engine is increased from about 60 psi to about 2,200 psi in the fuel rail by a camshaft-driven mechanical pump.

DFI Modes

The DFI has two different modes of operation: stratified and homogeneous. During cruising, the DFI engine is operating in the stratified or direct injection lean mode, and fuel is injected late in the compression cycle. Piston design and the injector spray pattern produce this stratified air-fuel charge by concentrating the fuel around the spark plug in the cylinder and making the fuel burn at a very lean 35:1 compared to 14.7:1 in a conventional engine. DFI is capable of saving 35 to 40 percent of the fuel consumed by a conventional gasoline engine at idle and cruising.

During acceleration, high speed, or heavy load, the fuel spray occurs in the intake stroke, like that of a conventional fuel injected engine. This mode is called homogeneous, and since the fuel in this mode is injected during the intake stroke, it is much like regular PFI.

The fuel injector can also be pulsed twice: once during a stratified charge mode and once during the homogenous mode as required for a smooth transition.

Typical DFI System

Direct fuel injection systems share many, if not most, components from the PFI systems, but have important differences.

The DFI injectors are of a special design so as to accommodate high fuel pressures and harsh conditions. Remember that the tips of the injectors are located in the combustion chamber itself, facing both high temperatures and pressures during their lifetime.

The majority of the DFI vehicles have some sort of variable valve timing control, turbocharging or supercharging, or both. Traditional camshaft and lifter arrangements are a compromise of idle quality, engine performance, and fuel mileage. With variable valve timing, the PCM can determine the ideal cam profile for a given engine operation. Variable valve timing allows engineers to have a perfect camshaft for idle, midrange, and full power. Variable valve timing often has the added benefit of being able to eliminate the EGR valve, the AIR system, and their associated controls and hardware.

There are several types of valve timing mechanisms, but most are hydraulically operated by actuators in the cylinder head. The center of the sprocket is attached to the camshaft. The outside of the sprocket is turned by the timing chain. The PCM controls pressurized oil through a pulse width modulated (PWM) oil control valve. If oil is directed to the left side of the oil cavity, the cam is advanced relative to the sprocket. If oil goes to the right, the cam is retarded.

In most models, the cams can be moved constantly within a 24-degree window. Both intake and exhaust cams can be fitted with variable valve timing, and some DOHC setups have variable valve timing on all four camshafts. More valve overlap is desirable at high speeds to increase scavenging of the combustion chamber, lowering the resistance to the flow of intake air and exhaust. Less overlap is desired at idle and for low-end torque.

Fuel Delivery System

The components of a typical gasoline delivery system are fuel tanks, fuel lines, fuel filters, and fuel pumps. Fuel is drawn from the fuel tank by an in-tank or chassis-mounted electric fuel pump. Before it reaches the injectors, the fuel passes through a filter that removes dirt and impurities. Fuel tanks are part of the evaporative emission controls (EVAP) that prevent gasoline vapors from leaving the tank. They also have an inlet filler tube and cap. Filler tube caps are non-venting and usually have some type of pressure-vacuum relief valve arrangement.

Some form of liquid/vapor separator is incorporated to stop liquid fuel or bubbles from reaching the vapor storage canister or the engine's crankcase. It can be located inside the tank, on the tank, in fuel vent lines, or near the fuel pump.

The fuel tank also houses a fuel sending unit that includes a pick-up tube and float-operated fuel gauge. The fuel tank pick-up tube is connected to the fuel pump by the fuel line. Most fuel pumps are combined with the sending unit called a fuel sender module.

Return-Line Style Systems

A fuel line pressure regulator, located on the fuel rail, maintains a constant fuel line pressure that may be as high as 65 psi in some PFI systems. This fuel pressure generates the spraying force needed to inject the fuel. Excess fuel not required by the engine returns to the fuel tank through a fuel return line.

Return-less Fuel Systems

Fuel systems that do not have a return line are called return-less fuel systems. In these systems, the fuel pressure regulator is located at the fuel pump/sender assembly instead of the fuel rail. This system has several advantages, two of which include no warm fuel being sent to the fuel tank (this raises the volatility of the fuel and makes a larger fuel charcoal canister necessary), and the elimination of the fuel return line. Additionally, less fuel is pumped through the system since fuel is not being recycled through the system continuously. Less fuel slowing through the system allows the use of a "lifetime" fuel filter installed inside the fuel sender/pump assembly. On DFI vehicles, the return-less fuel line is much preferred since the fuel pressure is very high, and would be hard to maintain over long lengths.

Many SFI engines took advantage of return-less fuel systems, but with DFI, the high fuel pressures needed to atomize the fuel (up to almost 2,200 psi) would be impractical to maintain from the tank to the injectors. DFI fuel systems will consist of a traditional in-tank pump plus a camshaft-driven (and in some cases PCM-controlled), mechanical, high pressure fuel pump. The PCM can control the output pressure of the high pressure fuel pump. This is done via pulse-width modulation of an inlet valve on the high pressure pump from a low of about 500 psi at idle to around 2,200 psi maximum. Different models will have different fuel pressures.

Fuel Lines and Fittings

Fuel lines can be made of either metal tubing or flexible nylon or synthetic rubber hose. The latter must be nonpermeable so that gasoline and gas vapors cannot evaporate through the hose. Similarly, vapor vent lines must be made of materials that resist attack by fuel vapors.

The fuel lines carry fuel from the fuel tank to the fuel pump, fuel filter, and fuel injection assembly. These lines are usually made of rigid metal, although some sections are constructed of rubber hose to allow for car vibrations.

Sections of fuel line are assembled together by fittings. Some of these fittings are a threaded-type fitting, while others are a quick-release design.

Fuel Filters

A strainer, located in the gasoline tank, is made of a finely woven fabric and is designed to prevent large contaminants from entering the fuel system where they could cause excessive fuel pump wear or plug fuel metering devices.

A fuel filter is connected in the fuel line between the fuel tank and the engine. Many of these filters are mounted under the vehicle and others are mounted in the engine compartment.

Fuel Pumps

An electric fuel pump is basically a small DC electric motor with an impeller mounted on the end of the motor's shaft. A pump cover is mounted over the impeller, and this cover contains inlet and discharge ports. When the armature and impeller rotate, fuel is moved from the tank to the inlet port, and the impeller grooves pick up the fuel and force it around the impeller cover and out the discharge port.

Fuel moves from the discharge port through the inside of the motor and out the check valve and outlet connection, which is connected via the fuel line to the fuel filter and underhood fuel system components. A pressure relief valve near the check valve opens if the fuel supply line is restricted and pump pressure becomes very high. When the relief valve opens, fuel is returned through this valve to the pump inlet. When the engine is shut off, the check valve prevents fuel from draining out of the fuel system into the fuel tank.

Electric fuel pump circuits vary depending on the vehicle make and year. In some circuits, an oil pressure switch is connected in series with the fuel pump. The oil pressure switch is wired in series to prevent engine damage from low oil pressure by turning off the fuel pump, making the engine stall.

Not only are fuel pumps controlled to prevent engine damage, they are also controlled to prevent fires during collisions. Various types of sensors are used that detect impact. When impact is sensed, the sensors open the fuel pump circuit. The PCM, in most vehicles, has ultimate control of the fuel pump and will only activate it when the conditions are safe for the passengers and the vehicle itself.

Emission Control Systems

Emission controls on cars and trucks have one purpose: to reduce the amount of pollutants and environmentally damaging substances released by the vehicles.

There are three main automotive pollutants: hydrocarbons (HC), carbon monoxide (CO), and oxides of nitrogen (NOx). Particulate emissions are also present in diesel engine exhaust. HC emissions are caused largely by unburned fuel from the combustion chambers. HC emissions can also originate from evaporative sources such as the gasoline tank. CO emissions are a by-product of the combustion process and result from incorrect air/fuel mixtures. NOx emissions are caused by nitrogen and oxygen uniting at cylinder temperatures above 2,500°F (1,371°C).

Three basic types of emission control systems are used in vehicles: evaporative control systems, precombustion, and post-combustion. The evaporative control system is a sealed system. It traps the fuel vapors (HC) that would normally escape from the fuel tank into the air.

Precombustion systems work to prevent emissions from being created in the engine, either during or before the combustion cycle. Anything that makes an engine more efficient can be categorized as a precombustion emission control. These precombustion control systems include the PCV and EGR systems.

Post-combustion control systems clean up the exhaust gases after the fuel has been burned. Secondary air or air injector systems put fresh air into the exhaust to reduce HC and CO to harmless water vapor and carbon dioxide by chemical (thermal) reaction with oxygen in the air. Catalytic converters help this process by reducing NOx, HC, and CO.

Evaporative Emission Control Systems

The fuel evaporative emission control system reduces the amount of raw fuel vapors that are emitted into the air from the fuel tank. These vapors must not be allowed to escape from the fuel system to the atmosphere.

Fuel vapors from the fuel tank are routed to and absorbed onto the surfaces of the canister's charcoal granules. When the vehicle is restarted, the vacuum draws vapors into the intake manifold to be burned in the engine. Canister purging varies widely with make and model. The PCM controls a relief and purge valve to clear the canister during conditions that are acceptable. The EVAP system is constantly monitored for leaks by the ECM through the EVAP monitor.

The onboard refueling vapor recovery system (ORVR) prevents fuel vapors from entering the atmosphere while the vehicle is being refueled.

PCV Systems

The PCV system is a precombustion emission control device. During the last part of the engine's combustion stroke, some unburned fuel and products of combustion leak past the engine's piston rings into the crankcase. This leakage into the engine crankcase is called blowby. Blowby must be removed from the engine before it condenses in the crankcase and reacts with the oil to form sludge. Sludge, if allowed to circulate with engine oil, corrodes and accelerates wear of pistons, piston rings, valves, bearings, and other internal working parts of the engine. Blowby gases must also be removed from the crankcase to prevent premature oil leaks. Because these gases enter the crankcase by the pressure formed during combustion, they pressurize the crankcase. The gases exert pressure on the oil pan gasket and crankshaft seals. If the pressure is not relieved, oil is eventually forced out of these seals.

Combustion gases that enter the crankcase are removed by a positive crankcase ventilation system, which uses engine vacuum to draw fresh air through the crankcase. This fresh air, which dissipates the harmful gases, enters through the air filter.

A PCV valve houses a tapered valve. When the engine is not running, a spring keeps the tapered valve seated against the valve housing. During idle or deceleration, the high intake manifold vacuum moves the tapered valve upward against the spring tension. Under this condition, there is a small opening between the tapered valve and the PCV valve housing. Since the engine is not under heavy load during idle or deceleration operation, blowby gases are minimal and the small PCV valve opening is adequate to move the blowby gases out of the crankcase.

Intake manifold vacuum is lower during part-throttle operation than during idle operation. Under this condition, the spring moves the tapered valve downward to increase the opening between this valve and the PCV valve housing.

Since engine load is higher at part-throttle operation than at idle operation, blowby gases are increased. The larger opening between the tapered valve and the PCV valve housing allows all the blowby gases to be drawn into the intake manifold.

When the engine is operating under heavy load conditions with a wide throttle opening, the decrease in intake manifold vacuum allows the spring to move the tapered valve further downward in the PCV valve. This action provides a larger opening between the tapered valve and the PCV valve housing. Since higher engine load results in more blowby gases, the larger PCV valve opening is necessary to allow these gases to flow through the valve into the intake manifold.

Some engines are equipped with a PCV system that does not use a PCV valve. Instead, the blowby gases are routed into the intake manifold through a fixed orifice tube. The basic system works the same as if it had a valve, except that the system is regulated only by the vacuum on the orifice. The size of the orifice limits the amount of blowby flow into the intake. The engine's air/fuel system is calibrated for this calibrated air leak. Since the action of the PCV allows unmetered air into the intake, the air/fuel system must be set for this amount of extra air.

Knock Sensor and Knock Sensor Module

Many engines with EFI have a knock sensor, or sensors. The knock sensors may be mounted in the block, cylinder head, or intake manifold. A piezoelectric sensing element is mounted in the knock sensor, and a resistor is connected parallel to this sensing element. When the engine detonates, a vibration occurs in the engine. The piezoelectric sensing element changes this vibration to an analog voltage, and this signal is sent to the knock sensor module.

The knock sensor module changes the analog voltage signal to a digital voltage signal and sends this signal to the PCM. When the PCM receives this signal, it reduces the spark advance to prevent detonation.

EGR Systems

The EGR system dilutes the air/fuel mixture with controlled amounts of exhaust gas. Since exhaust gas does not burn, this reduces the peak combustion temperatures. At lower combustion temperatures, very little of the nitrogen in the air combines with oxygen to form NOx. Most of the nitrogen is simply carried out with the exhaust gases. For driveability, it is desirable to have the EGR valve opening (and the amount of gas flow) proportional to the throttle opening. Driveability is also improved by shutting off the EGR when the engine is started up cold, at idle, and at full throttle. Since the NOx control requirements vary on different engines, there are several different systems, with various controls to provide these functions.

Typically, the EGR valve is an electrically operated flow control valve. On most systems, it is attached to the intake manifold. A small exhaust crossover pipe admits exhaust gases to the inlet port of the EGR valve. Opening the EGR valve allows exhaust gases to flow through the valve, where it mixes with the air/fuel mixture. The effect is to dilute or lean-out the mixture so that it still burns completely but with a reduction in combustion chamber temperatures.

On some engines, the exhaust gas from the EGR system is distributed through passages in the cylinder heads and distribution plates to each intake port. Since the exhaust gas from the EGR system is distributed equally to each cylinder, smoother engine operation results.

The exhaust gas flow rate depends on commands from the ECM. There are two types of EGR valves in common use. These are either vacuum or electrically operated units.

The vacuum-operated EGR valve uses an electronically operated vacuum regulator valve. The EGR opening can be varied by pulsing the vacuum regulator for varying lengths of time, called pulse width modulation. On these EGR valves, the functioning of the valve is verified by a differential pressure sensor in the exhaust supply passage.

There are two types of electrically operated types of valves: digital and linear. A digital EGR valve contains up to three electric solenoids that are operated directly by the PCM. Each solenoid contains a movable plunger with a tapered tip that seats in an orifice. When any solenoid is energized, the plunger is lifted and exhaust gas is allowed to recirculate through the orifice into the intake manifold. The solenoids and orifices are different sizes. The PCM can operate one, two, or three solenoids to supply the amount of exhaust recirculation required to provide optimum control of NOx emissions.

The linear EGR valve contains a single electric solenoid that is operated by the PCM. A tapered pintle is positioned on the end of the solenoid plunger. When the solenoid is energized, the plunger and tapered valve are lifted, and exhaust gas is allowed to recirculate into the intake manifold. The EGR valve contains an EGR valve position (EVP) sensor, which is a linear potentiometer. The signal from this sensor varies from approximately 1 V with the EGR valve closed to 4.5 V with the valve wide open.

The PCM pulses the EGR solenoid winding on and off with pulse width modulation to provide accurate control of the plunger and EGR flow. The EVP sensor acts as a feedback signal to the PCM to inform the PCM if the commanded valve position was achieved.

Catalytic Converters

Post-combustion emission control devices clean up the exhaust after the fuel has been burned but before the gases exit the vehicle's tailpipe. An excellent example of this is the catalytic converter. A converter is one of the most effective emission control devices on a vehicle for reducing HC, CO, and NOx.

A catalytic converter contains a ceramic element coated with a catalyst. A catalyst is something that causes a chemical reaction without being part of the reaction. A catalytic converter causes a chemical change to take place in the passing exhaust gases. Most of the harmful gases are changed to harmless gases.

Three different materials are used as the catalyst in automotive converters: platinum, palladium, and rhodium. Platinum and palladium are the oxidizing elements of a converter. When HC and CO are exposed to heated surfaces covered with platinum and palladium, a chemical reaction takes place. The HC and CO are combined with oxygen to become H_2O and CO_2. Rhodium is a reducing catalyst. When NOx is exposed to hot rhodium, oxygen is removed and NOx becomes just N. The removal of oxygen is called reduction, which is why rhodium is a reducing catalyst.

A catalytic converter that contains all three catalysts and reduces HC, CO, and NOx is called a three-way converter. Three-way converters have the oxidizing catalysts in part of the container and the reducing catalyst in the other.

Air Injection Systems

One of the earliest methods used to reduce the amount of hydrocarbons and carbon monoxide in the exhaust was by forcing fresh air into the exhaust system after combustion. This additional fresh air causes further oxidation and burning of the unburned hydrocarbons and carbon monoxide. Oxygen in the air combines with the HC and CO to continue the burning that reduces the HC and CO concentrations. This allows them to oxidize and produce harmless water vapor and carbon dioxide. The modern secondary air pump is used during cold starts, for about the first 90 seconds of operation. The purpose is mainly to clean up cold start emissions. The secondary air monitor is used to determine the functionality of the system. The ECM can command the air pump to turn on for a few seconds following engine warmup. The oxygen sensor should show a very lean mixture under these conditions.

The typical air injection system includes the following:

- An electronically operated air pump
- A control valve to direct the flow of air and prevent the intrusion of exhaust gas
- Hoses and nozzles that are used to distribute and inject the air from the pump

Computer-Controlled Engine Systems

Because all manufacturers have continually updated, expanded, and improved their computerized control systems, there are hundreds of different domestic and import systems on the road. In a computerized engine control system, emission levels, fuel consumption, driveability, and durability are carefully balanced to achieve maximum results with minimum waste. Some of the things engine control systems are designed to do are:

- Air/fuel ratios are held as closely to 14.7 to 1 as possible, allowing maximum catalytic converter efficiency and minimizing fuel consumption.
- Emission control devices, such as EGR valve, carbon canister, and air pump, are operated at predetermined times to increase efficiency.
- The engine is operated as efficiently as possible when it is cold and is warmed up

rapidly, reducing unburned hydrocarbon emissions and engine wear due to raw gas washing oil from the piston rings and getting into the crankcase to form sludge and varnish.

- Ignition timing is advanced as much as possible under all conditions.
- Timing and air/fuel ratios are precisely controlled under all operating conditions.
- Control loop operation enables the engine to make rapid changes to match changes in engine temperature, load, and speed.

In a PCM, RAM is used to store data collected by the sensors, the results of calculations, and other information that is constantly changing during engine operation. Information in volatile RAM is erased when the ignition is turned off or when the power to the computer is disconnected. Nonvolatile RAM does not lose its data if its power source is disconnected.

The computer's permanent memory is stored in read only memory (ROM) or programmable read only memory (PROM). Like non-volatile RAM, ROM and PROM are not erased when the power source is disconnected. ROM and PROM are used to store computer control system strategy and look-up tables. PROM normally contains the specific information about the vehicle it is installed in.

System adaptive strategy is a plan, created by engine designers and calibration engineers, for the timing and control of computer-controlled systems. In designing the best strategies, it is necessary to look at all the possible conditions an engine may encounter. It is then determined how the system should respond to these conditions.

The look-up tables (sometimes called maps) contain calibrations and specifications. Look-up tables indicate how an engine should perform. For example, information (a reading of 20 in. Hg) is received from the manifold absolute pressure (MAP) sensor. This information, plus information from the engine speed sensor, is compared to a table for spark advance. This table tells the computer what the spark advance should be for that throttle position and engine speed. The computer then modifies the spark advance.

When making decisions, the PCM is constantly referring to three sources of information: the look-up tables, system strategy, and the input from sensors. By comparing information from these sources, the computer makes informed decisions.

The computer has adaptive strategy capabilities, it can actually learn from past experience. For example, the normal voltage signals from the TP sensor to the PCM range from 0.6 to 4.5 volts. If a 0.2-volt signal is received, the PCM may regard this signal as the result of a worn TP sensor and assign this lower voltage to the normal low voltage signal. In other words, the PCM will add 0.4 volts to the 0.2 volts it received. All future signals from the various throttle positions will also have 0.4 volt added to the signal. Doing this calculation adjusts for the worn TP sensor and ensures that the engine will operate normally. If the input from a sensor is erratic or considerably out of range, the PCM may totally ignore the input.

When a computer has adaptive learning, a short learning period is necessary after the battery has been disconnected, when a computer has been disconnected or replaced, or when the vehicle is new. During this learning period, the engine may surge, idle fast, or have a loss of power. The average learning period lasts for five miles of driving.

Most adaptive strategies have two parts: Short Term Fuel Trim and Long Term Fuel Trim. Short-term strategies are those immediately enacted by the computer to overcome a change in operation. These changes are temporary. Long-term strategies are based on the feedback about the short-term strategies. These changes are more permanent.

OBDII

OBD II (On-Board Diagnostic System, Generation 2) has monitors that evaluate such things as the catalytic converters, engine misfire detection, evaporative system, secondary air system, and EGR system flow rate. These monitors detect problems that would affect emissions levels. Also, all OBDII systems provide a serial data stream of twenty basic data parameters and have common diagnostic trouble codes known as generic trouble codes. Manufacturer-specific codes are also used by automakers.

OBD II PCMs have electrically erasable programmable read-only memory (EEPROM) to store data without the need for a continuing source of electrical power. It is an integrated circuit that contains the program used by the PCM to provide power train control. When a modification to the PCM operating strategy is required, the EEPROM may be reprogrammed through the DLC using computer software.

OBD II systems monitor the effectiveness of the major emission control systems, and anything else that may affect emissions. This is done by the PCM, which runs certain tests on various subsystems of the engine management system. If one or more monitored systems are found to have a malfunction, the MIL will be illuminated to alert the driver of a problem and a diagnostic trouble code is stored in the PCM. Some monitors run continuously, while others will run only when certain operating conditions are present. These conditions are called the enable criteria.

Typical Sensors

To monitor engine conditions, the computer uses a variety of sensors. All sensors perform the same basic function. They detect a mechanical condition (movement or position), chemical state, or temperature condition and change it into an electrical signal that can be used by the PCM to make decisions. The following represent the most commonly used sensors. It should be noted that these sensors are shared between the systems. For example, MAP sensor input is used to control fuel, ignition, EGR, emission system airflow, air intake, and idle speed systems.

Engine Coolant Temperature (ECT) Sensor

The ECT sensor is very important. Its input is used to regulate many engine functions, such as activating and deactivating the EGR, canister purge, and TCC systems and controlling the open- and closed-loop feedback modes of the system.

The ECT sensor is usually located on the intake manifold. The sensor screws into a water jacket.

Engine Position Sensors

Engine position sensors tell the computer the speed of the engine and when the piston in each cylinder reaches top dead center (TDC). This input is used to set ignition timing and fuel injection delivery. Several distinct types of engine position sensors are used, but all communicate with the computer by generating a voltage signal. The sensor does this using a Hall-effect switch or magnetic pulse generator. The engine position sensor may be called a distributor pick-up coil, crankshaft or camshaft position sensor, or a profile ignition pick-up sensor.

EGR Valve Position Sensor

Car manufacturers use a variety of sensors or switches to determine when the EGR valve is open. This information is used to adjust the air/fuel mixture. The exhaust gases introduced by the EGR valve into the intake manifold reduce the available oxygen and thus less fuel is needed in order to maintain low HC levels in the exhaust. Most EGR valve position sensors are linear potentiometers mounted on top of the EGR valve. When the EGR valve opens, the potentiometer stem moves upward and a higher voltage signal is sent to the PCM.

Feedback Pressure EGR Sensor

The feedback pressure EGR sensor used in some vehicles is a pressure-sensing voltage divider (functions as a potentiometer) similar to the ones used as MAP and BARO sensors. It senses exhaust pressure in a chamber just under the EGR valve. This pressure causes the sensor to vary its output voltage signal to the computer. When the EGR valve is closed, the pressure in the sensing chamber is equal to exhaust pressure. When the EGR valve opens, pressure in this chamber drops because of the restricting orifice that lets exhaust into the sensing chamber from the exhaust system. The more the valve opens, the more the pressure drops. The feedback pressure EGR sensor's voltage signal tells the computer how far the EGR valve is open. The computer uses this information to fine tune its control of the electronic vacuum regulator, which controls vacuum to the EGR valve. This information also allows the computer to more accurately control air/fuel ratios and ignition timing.

Intake Air Temperature (IAT) Sensor

Also referred to as the air change temperature sensor, this sensor is a thermistor. Its resistance decreases as manifold air temperature increases and increases as manifold air temperature decreases. The PCM measures the voltage drop across the sensor and uses this input to help calculate fuel delivery. Cold intake air is denser; therefore, a richer air-fuel ratio is required. When the

IAT signal indicates colder intake air temperature, the PCM provides a richer air-fuel ratio. The input from this sensor may also be used to control the preheated air and early fuel evaporation systems.

Manifold Absolute Pressure (MAP) Sensor

The function of a MAP sensor is to sense air pressure or vacuum in the intake manifold. The computer uses this input as an indication of engine load to adjust the air/fuel mixture and spark timing. The MAP sensor reads vacuum and pressure through a hose connected to the intake manifold. A pressure-sensitive ceramic or silicon element and electronic circuit in the sensor generates a voltage signal that changes in direct proportion to pressure.

MAP sensors should not be confused with vacuum sensors or barometric pressure sensors. While a vacuum sensor reads the difference between manifold vacuum and atmospheric pressure, a MAP sensor measures manifold air pressure against a precalibrated absolute pressure. Because it bases its readings on preset absolute pressure, MAP sensor readings are not adversely altered by changes in operating altitudes or barometric pressure.

Mass Airflow (MAF) Sensor

This sensor measures the flow of air entering the engine. This measurement of airflow is a reflection of engine load (throttle opening and air volume). It is similar to the relationship of engine load to MAP or vacuum sensor signal. Since there are several types of MAFs, check the service manual for the one used.

In the heated resistor-type, a heated resistor is mounted in the center of the air passage. When the ignition switch is turned on, voltage is applied to the sensor's module and the module allows enough current flow through the resistor to maintain a specific resistor temperature. If a cold engine is accelerated suddenly, the rush of cool air tries to cool the resistor. Under this condition, more current is needed to maintain the temperature of the resistor. The sensor's module sends a signal to the PCM indicating the amount of current needed to maintain the temperature. When the PCM receives a signal indicating more airflow or higher current, it allows for more fuel to be delivered to the cylinders.

In a hot-wire-type MAF, a hot wire is positioned in the air stream through the sensor, and an ambient temperature sensor wire is located beside the hot wire. This ambient sensor wire, sometimes called the cold wire, senses intake air temperature. When the ignition switch is turned on, the module in the MAF sensor allows enough current flow through the hot wire to allow it to maintain a specific number of degrees above the ambient temperature measured by the cold wire. Like the heated resistor-type MAF, if the engine is suddenly accelerated, the rush of cold air tends to cool the hot wire. As a result, more current is needed to maintain the desired temperature and a signal is sent to the PCM indicating how much current is required. Based on this input, the PCM adds or subtracts fuel in an attempt to achieve the correct air-fuel ratio for the conditions.

Oxygen Sensors (O_2S)

The exhaust gas oxygen sensor is the key sensor in the closed-loop mode. Its input is used by the computer to maintain a balanced air/fuel mixture. The O_2S is threaded into the exhaust manifold or into the exhaust pipe near the engine.

One type of oxygen sensor, made with a zirconium dioxide element, generates a voltage signal proportional to the amount of oxygen in the exhaust gas. It compares the oxygen content in the exhaust gas with the oxygen content of the outside air. As the amount of unburned oxygen in the exhaust gas increases, the voltage output of the sensor drops. Sensor output ranges from 0.1 volt (lean) to 0.9 volt (rich). A perfectly balanced air/fuel mixture of 14.7:1 produces an output of around 0.5 volt. When the sensor reading is lean, the computer enriches the air/fuel mixture to the engine. When the sensor reading is rich, the computer leans the air/fuel mixture.

Because the oxygen sensor must be hot to operate, all late-model engines use heated oxygen sensors (HO_2S). These sensors have an internal heating element that allows the sensor to reach operating temperature more quickly and to maintain its temperature during periods of idling or low engine load.

Air Fuel Ratio Sensor (Wide-Band Oxygen Sensor)

Although the oxygen sensor and the AF ratio sensor both measure oxygen in the exhaust, there are some fundamental differences in their operation. First, the air-fuel sensor operates at 1,200°F (650°C), about twice as hot as a conventional oxygen sensor. In addition, the A/F sensor changes amperage and voltage to indicate the A/F ratio. When the mixture is lean, the current is generated in a positive direction with a voltage above 3.3 volts. When the mixture is rich, the current is generated in a negative direction with a voltage below 3.3 volts. At 14.7:1, no current is produced.

The current flow originates inside the ECM and is in the range of 1.5 mA or less. Because it is difficult to measure with conventional voltmeters or oscilloscopes, most manufacturers recommend using a scan tool only for diagnosis of these sensors. The A/F sensor does not toggle rich to lean in operation like the oxygen sensor. The A/F sensor can not only indicate rich and lean like the oxygen sensor, but determine just how rich or lean the mixture is by varying the current direction and voltage. Thus, the computer can make better decisions on fuel management since it has more accurate information.

When looking at a scan tool reading for the AF sensor, remember that early OBD II regulations require oxygen sensors to read out between 0 volt and 1 volt, so the scan tool may convert the signal to correspond to the conventional sensor, or the reading is translated into a lambda reading. A lambda reading greater than 1 indicates a rich condition, while a lambda reading less than 1 indicates a lean condition.

The PCM runs a fixed injection amount to monitor the AF sensor. This leads to an accurate diagnostic on the sensor itself, which is performed by the ECM during the monitor for the AF Sensor. Some manufacturers call the AF sensor an oxygen sensor. The technician can tell the difference by looking at the wiring connector. Most conventional oxygen sensors have about four wires in their connector, while the AF sensor generally has about six or seven depending on the manufacturer.

Throttle Position (TP) Sensor

Engines with electronic fuel injection or feedback carburetors use a TP sensor to inform the computer about the rate of throttle opening and the relative throttle position. The TP sensor contains a potentiometer with a pointer that is rotated by the throttle shaft. As the throttle shaft moves, the pointer moves to a new location on the resistor in the potentiometer. The return voltage signal tells the PCM how much the throttle plates are open. As resistance readings tell the computer that the throttle is opening, it enriches the air/fuel mixture to maintain the proper air-fuel ratio. A separate idle switch or wide open throttle (WOT) switch may also be used to signal the computer when these throttle positions exist.

The initial setting of the sensor is critical. The voltage signal the computer receives is referenced to this setting. Many service manuals list the initial TP sensor setting to the nearest one-hundredth volt, a clear indication of the importance of this setting.

Vehicle Speed Sensor (VSS)

This sensor tells the computer the vehicle's speed in miles per hour, which is fed to the instrument cluster for the speedometer and the odometer. This input also helps determine transmission shifting points, cruise control, EGR flow, and canister purge, among others. The VSS is mounted in the transmission/transaxle opening or on the transfer case.

Computer Outputs and Actuators

Once the PCM's programming instructs that a correction or adjustment must be made in the controlled system, an output signal is sent to a control device or actuator. These actuators, which are solenoids, switches, relays, or motors, physically act or carry out the command sent by the PCM.

Actuators are electromechanical devices that convert an electrical current into mechanical action. This mechanical action can then be used to open and close valves, control vacuum to other components, or open and close switches. When the PCM receives an input signal indicating a change in one or more of the operating conditions, the PCM determines the best strategy for handling the conditions. The PCM then controls a set of actuators to achieve a desired effect or strategy goal. In order for the computer to control an actuator, it must rely on an output driver.

The circuit driver usually applies the ground circuit of the actuator. The ground can be applied steadily if the actuator must be activated for a selected amount of time. Or the ground can be pulsed to activate the actuator in pulses. Output drivers are transistors or groups of transistors that control the actuators. These drivers operate by the digital commands from the PCM. If an actuator can't be controlled digitally, the output signal must pass through an A/D converter before flowing to the actuator. The major actuators in a computer-controlled engine include the following components.

- *Air Management Solenoids*—Secondary air switching solenoids control the flow of air from the air pump to either the exhaust manifold or to the atmosphere.
- *Evaporative Emission (EVAP) Canister Purge and Vent Valves*—These valves are controlled by solenoids. The purge valve controls when stored fuel vapors in the canister are drawn into the engine and burned. The vent valve is used to close off the system for leak testing.
- *EGR Flow Solenoids*—EGR flow may be controlled by electronically controlled solenoids. Some EGR systems use solenoid valves to supply manifold vacuum to the EGR valve when EGR is required. Some systems are operated by a stepper motor.
- *Fuel Injectors*—These solenoid valves deliver the fuel spray in fuel-injected systems.
- *Throttle Actuator Control TAC*—A stepper motor is used to electrically actuate the throttle plate. Older systems used a stepper motor to control the idle air control.
- *Motors and Lights*—Using electrical relays, the computer is used to trigger the operation of electric motors such as the fuel pump, or various warning light or display circuits.
- *Automatic Transmission Shifting*—Computer-controlled solenoids are also used in the operation of automatic transmission shift mechanisms.

System Operation

Control loops are the cycles by which a process can be controlled by information received from input sensors, ROM, computer processing, and output of specific commands to control actuator devices.

The basic purpose of all computerized engine control loops is the same: to create an ideal air-fuel ratio, which allows the catalytic converter to operate at maximum efficiency, while giving the best mileage and performance possible and protecting the engine.

The closed loop mode is basically the same for any automotive system. Sensor inputs are sent to the computer, the computer compares the values to its programs, and then sends commands to the output devices. The output devices adjust timing, air-fuel ratio, and emission control operation. The resulting engine operation affects the sensors, which send new messages to the computer, completing the cycle of operation. The complete cycle is called a closed loop.

Closed control loops are often referred to as feedback systems. This means that the sensors provide constant information, or feedback, on what is taking place in the engine. This allows the computer to make constant decisions and changes to output commands.

When the engine is cold, most electronic engine controls go into open loop mode. In this mode, the control loop is not a complete cycle because the computer does not react to feedback information. Instead, the computer makes decisions based on preprogrammed information that allows it to make basic ignition or air/fuel settings and to disregard sensor inputs. The open loop mode is activated when a signal from the temperature sensor indicates that the engine temperature is too low for gasoline to properly vaporize and burn in the cylinders. Systems with oxygen sensors may also go into the open loop mode while idling, or at any time that the oxygen sensor cools off enough to stop sending a signal, and at wide open throttle.

Multiplexing

Multiplexing is an in-vehicle networking system used to transfer data between electronic modules through a serial data bus. Serial data is electronically coded information that is transmitted by one computer and received and displayed by another computer. Serial data is information that is digitally coded and transmitted in a series of data bits. The data transmission rate is referred to as the baud rate. Baud rate refers to the number of data bits that can be transmitted in a second.

With multiplexing, fewer dedicated wires are required for each function, and this reduces the size of the wiring harness. Using a serial data bus reduces the number of wires by combining the signals on a single wire through time division multiplexing. Information is sent to individual control modules that control each function, such as engine controls, transmission controls, antilock braking, and many different accessories.

Multiplexing also eliminates the need for redundant sensors because the data from one sensor is available to all electronic modules. Multiplexing also allows for greater vehicle content flexibility because functions can be added through software changes, rather than adding another module or modifying an existing one.

The common multiplex system is called the CAN or Controller Area Network. It is used to interconnect a network of electronic control modules.

Hybrid Systems

A hybrid electric vehicle (HEV) uses one or more electric motors and an engine to propel the vehicle. Depending on the design of the system, the engine may move the vehicle by itself, assist the electric motor while it is moving the vehicle, or it may drive a generator to charge the vehicle's batteries. The electric motor may power the vehicle by itself or assist the engine while it is propelling the vehicle. Many hybrids rely exclusively on the electric motor(s) during slow speed operation, the engine at higher speeds, and both during some certain driving conditions. Complex electronic controls monitor the operation of the vehicle, based on the current operating conditions, electronics control the engine, electric motor, and generator.

A hybrid's electric motor is powered by high-voltage batteries, which are recharged by a generator driven by the engine and through regenerative braking. Regenerative braking is the process by which a vehicle's kinetic energy can be captured while it is decelerating and braking. The electric drive motors become generators driven by the vehicle's wheels. These generators take the kinetic energy, or the energy of the moving vehicle, and change it into energy that charges the batteries. The magnetic forces inside the generator cause the drive wheels to slow down. A conventional brake system brings the vehicle to a safe stop.

The engines used in hybrids are specially designed for the vehicle and electric assist. Therefore, they can operate more efficiently; resulting in very good fuel economy and very low tailpipe emissions. HEVs can provide the same performance, if not better, as a comparable vehicle equipped with a larger engine.

There are primarily two types of hybrids: the parallel and the series designs. A parallel hybrid electric vehicle uses either the electric motor or the gas engine to propel the vehicle, or both. The engine in a true series hybrid electric vehicle is used only to drive the generator that keeps the batteries charged. The vehicle is powered only by the electric motor(s). Most current HEVs are considered to have a series/parallel configuration because they have the features of both designs.

Although most current hybrids are focused on fuel economy, the same ideas can be used to create high-performance vehicles. Hybrid technology is also influencing off-the-road performance. By using individual motors at the front and rear drive axles, additional power can be applied to certain drive wheels, when needed.

SAFETY

In an automotive repair shop, there is great potential for serious accidents, simply because of the nature of the business and the equipment used. When people are careless, the automotive repair industry can be one of the most dangerous occupations. But the chances of your being injured while working on a car are close to nil if you learn to work safely and use common sense. Safety is the responsibility of everyone in the shop.

Personal Protection

Some procedures, such as grinding, result in tiny particles of metal and dust that are thrown off at very high speeds. These metal and dirt particles can easily get into your eyes, causing scratches or cuts on your eyeball. Pressurized gases and liquids escaping a ruptured hose or hose-fitting can spray a great distance. If these chemicals get into your eyes, they can cause blindness. Dirt and sharp bits of corroded metal can easily fall down into your eyes while you are working under a vehicle.

Eye protection should be worn whenever you are exposed to these risks. To be safe, you should wear safety glasses whenever you are working in the shop. Some procedures may require that you wear other eye protection in addition to safety glasses. For example, when cleaning parts with a pressurized spray, you should wear a face shield. The face shield not only gives added protection to your eyes but also protects the rest of your face.

If chemicals such as battery acid, fuel, or solvents get into your eyes, flush them continuously with clean water. Have someone call a doctor and get medical help immediately.

Your clothing should be well fitted and comfortable but made of strong material. Loose, baggy clothing can easily get caught in moving parts and machinery. Some technicians prefer to wear coveralls or shop coats to protect their personal clothing. Your work clothing should offer you some protection but should not restrict your movement.

Long hair and loose, hanging jewelry can create the same type of hazard as loose-fitting clothing. They can get caught in moving engine parts and machinery. If you have long hair, tie it back or tuck it under a cap.

Never wear rings, watches, bracelets, and neck chains. These can easily get caught in moving parts and cause serious injury.

Always wear shoes or boots of leather or similar material with non-slip soles. Steel-tipped safety shoes can give added protection to your feet. Jogging or basketball shoes, street shoes, and sandals are inappropriate in the shop.

Good hand protection is often overlooked. A scrape, cut, or burn can limit your effectiveness at work for many days. A well-fitted pair of heavy work gloves should be worn during operations such as grinding and welding or when handling hot components. Always wear approved rubber gloves when handling strong and dangerous caustic chemicals.

Many technicians wear thin, surgical-type latex gloves whenever they are working on vehicles. These offer little protection against cuts but do offer protection against disease and grease buildup under and around your fingernails. These gloves are comfortable and are quite inexpensive.

Accidents can be prevented simply by the way you act. The following are some guidelines to follow while working in a shop. This list does not include everything you should or shouldn't do; it merely presents some things for you to think about.

- Never smoke while working on a vehicle or while working with any machine in the shop.
- Playing around is not fun when it sends someone to the hospital.
- To prevent serious burns, keep your skin away from hot metal parts such as the radiator, exhaust manifold, tailpipe, catalytic converter, and muffler.
- Always disconnect electric engine cooling fans when working around the radiator. Many of these will turn on without warning and can easily chop off a finger or hand. Make sure you reconnect the fan after you have completed your repairs.
- When working with a hydraulic press, make sure the pressure is applied in a safe manner. It is generally wise to stand to the side when operating the press.
- Properly store all parts and tools by putting them away in a place where people will not trip over them. This practice not only cuts down on injuries, but also reduces time wasted looking for a misplaced part or tool.

Work Area Safety

Your entire work area should be kept clean and safe. Any oil, coolant, or grease on the floor can make it slippery. To clean up oil, use commercial oil absorbent. Keep all water off the floor. Water not only makes smooth floors slippery, but it is also dangerous as a conductor of electricity. Aisles and walkways should be kept clean and wide enough to allow easy movement. Make sure the work areas around machines are large enough so that the machines can be safely operated.

Gasoline is a highly flammable volatile liquid. Something that is flammable catches fire and burns easily. A volatile liquid is one that vaporizes very quickly. Flammable volatile liquids are potential firebombs. Always keep gasoline or diesel fuel in an approved safety can and never use gasoline to clean your hands or tools.

Handle all solvents (or any liquids) with care to avoid spillage. Keep all solvent containers closed, except when pouring. Proper ventilation is very important in areas where volatile solvents and chemicals are used. Solvents and other combustible materials must be stored in approved and designated storage cabinets or rooms with adequate ventilation. Never light matches or smoke near flammable solvents and chemicals, including battery acids.

Oily rags should also be stored in an approved metal container. When these oily, greasy, or paint-soaked rags are left lying about or are not stored properly, they can cause spontaneous combustion. Spontaneous combustion results in a fire that starts by itself, without a match.

Disconnecting the vehicle's battery before working on the electrical system, or before welding, can prevent fires caused by a vehicle's electrical system. To disconnect the battery, remove the negative or ground cable from the battery and position it away from the battery.

Know where all of the shop's fire extinguishers are located. Fire extinguishers are clearly labeled as to what type they are and what types of fire they should be used on. Make sure you use the correct type of extinguisher for the type of fire you are dealing with. A multipurpose dry chemical fire extinguisher will put out ordinary combustibles, flammable liquids, and electrical fires. Never put water on a gasoline fire because that will just cause the fire to spread. The proper fire extinguisher will smother the flames.

During a fire, never open doors or windows unless it is absolutely necessary; the extra draft will only make the fire worse. Make sure the fire department is contacted before or during your attempt to extinguish a fire.

Battery Safety

The potential dangers caused by the sulfuric acid in the electrolyte and the explosive gases generated during battery charging require that battery service and troubleshooting be conducted under absolutely safe working conditions. Always wear safety glasses or goggles when working with batteries no matter how small the job.

Sulfuric acid can also cause severe skin burns. If electrolyte contacts your skin or eyes, flush the area with water for several minutes. When eye contact occurs, force your eyelid open. Always have a bottle of neutralizing eyewash on hand and flush the affected areas with it. Do not rub your eyes or skin. Receive prompt medical attention if electrolyte contacts your skin or eyes. Call a doctor immediately.

When a battery is charging or discharging, it gives off quantities of highly explosive hydrogen gas. Some hydrogen gas is present in the battery at all times. Any flame or spark can ignite this gas, causing the battery to explode violently, propelling the vent caps at a high velocity and spraying acid over a wide area. To prevent this dangerous situation, take these precautions:

- Never smoke near the top of a battery and never use a lighter or match as a flashlight.
- Remove wristwatches and rings before servicing any part of the electrical system. This helps to prevent the possibility of electrical arcing and burns.
- Even sealed, maintenance-free batteries have vents and can produce dangerous quantities of hydrogen if severely overcharged.
- When removing a battery from a vehicle, always disconnect the battery ground cable first. When installing a battery, connect the ground cable last.
- Always disconnect the battery's ground cable when working on the electrical system or engine. This prevents sparks from short circuits and prevents accidental starting of the engine.
- Always operate charging equipment in well-ventilated areas. A battery that has been overworked should be allowed to cool down. Let air circulate around it before attempting to jump-start the vehicle. Most batteries have flame arresters in the caps to help prevent explosions, so make sure that the caps are tightly in place.
- Never connect or disconnect charger leads when the charger is turned on. This generates a dangerous spark.
- Never lay metal tools or other objects on the battery because a short circuit across the terminals can result.
- Always disconnect the battery ground cable before fast-charging the battery on the vehicle. Improper connection of charger cables to the battery can reverse the current flow and damage the AC generator.
- Never attempt to use a fast charger as a boost to start the engine.
- As a battery gets closer to being fully discharged, the acidity of the electrolyte is reduced, and the electrolyte starts to behave more like pure water. A dead battery may freeze at temperatures near zero degrees Fahrenheit. Never try to charge a battery that has ice in the cells.

Passing current through a frozen battery can cause it to rupture or explode. If ice or slush is visible or the electrolyte level cannot be seen, allow the battery to thaw at room temperature before servicing. Do not take chances with sealed batteries. If there is any doubt, allow them to warm to room temperature before servicing.

- As batteries get old, especially in warm climates and especially with lead-calcium cells, the grids start to grow. The chemistry is rather involved, but the point is that plates can grow to the point where they touch, producing a shorted cell.
- Always use a battery carrier or lifting strap to make moving and handling batteries easier and safer.
- Acid from the battery damages a vehicle's paint and metal surfaces and harms shop equipment. Neutralize any electrolyte spills during servicing.

Working Safely on High-Voltage Systems

Electric drive vehicles (battery-operated, hybrid, and fuel cell electric vehicles) have high-voltage electrical systems (from 42 volts to 650 volts). These high voltages can kill you! Fortunately, most high-voltage circuits are identifiable by size and color. The cables have thicker insulation and are typically colored orange. The connectors are also colored orange. On some vehicles, the high-voltage cables are enclosed in an orange shielding or casing—again, the orange indicates high voltage. In addition, the high-voltage battery pack and most high-voltage components have "High Voltage" caution labels. Be careful not to touch these wires and parts.

Wear insulating gloves, commonly called "lineman's gloves," when working on or around the high-voltage system. These gloves must be class "0" rubber insulating gloves, rated at 1000-volts. Also, to protect the integrity of the insulating gloves, as well as you, wear leather gloves over the insulating gloves while doing a service.

Make sure they have no tears, holes or cracks and that they are dry. Electrons are very small and can enter through the smallest of holes in your gloves. The integrity of the gloves should be checked before using them. To check the condition of the gloves, blow enough air into each one so they balloon out. Then fold the open end over to seal the air in. Continue to slowly fold that end of the glove toward the fingers. This will compress the air. If the glove continues to balloon as the air is compressed, it has no leaks. If any air leaks out, the glove should be discarded. All gloves, new and old, should be checked before they are used.

There are other safety precautions that should always be adhered to when working on an electric drive vehicle:

- Always adhere to the safety guidelines given by the vehicle's manufacture.
- Obtain the necessary training before working on these vehicles.
- Be sure to perform each repair operation following the test procedures defined by the manufacturer.
- Disable or disconnect the high-voltage system before performing services to those systems. Do this according to the procedures given by the manufacturer.
- Anytime the engine is running in a hybrid vehicle, the generator is producing high-voltage and care must be taken to prevent being shocked.
- Before doing any service to an electric drive vehicle, make sure the power to the electric motor is disconnected or disabled.
- Systems may have a high-voltage capacitor that must be discharged after the high-voltage system has been isolated. Make sure to wait the prescribed amount of time (normally about 10 minutes) before working on or around the high-voltage system.
- After removing a high-voltage cable, cover the terminal with vinyl electrical tape.
- Always use insulated tools.
- Alert other technicians that you are working on the high-voltage systems with a warning sign such as "High Voltage Work: Do Not Touch."
- Always install the correct type of circuit protection device into a high-voltage circuit.

- Many electric motors have a strong permanent magnet in them; individuals with a pacemaker should not handle these parts.
- When an electric drive vehicle needs to be towed into the shop for repairs, make sure it is not towed on its drive wheels. Doing this will drive the generator(s), which can overcharge the batteries and cause them to explode. Always tow these vehicles with the drive wheels off the ground or move them on a flat bed.

Air Bag Safety

When service is performed on any air bag system component, always disconnect the negative battery cable, isolate the cable end, and wait for the amount of time specified by the vehicle manufacturer before proceeding with the necessary diagnosis or service. The average waiting period is two minutes, but some vehicle manufacturers specify up to ten minutes. Failure to observe this precaution may cause accidental air bag deployment and personal injury.

Replacement air bag system parts must have the same part number as the original part. Replacement parts of lesser or questionable quality must not be used. Improper or inferior components may result in inappropriate air bag deployment and injury to vehicle occupants.

Do not strike or jar a sensor or an air bag system diagnostic monitor (ASDM). This may cause air bag deployment or make the sensor inoperative. Accidental air bag deployment may cause personal injury, and an inoperative sensor may result in air bag deployment failure, causing personal injury to vehicle occupants.

All sensors and mounting brackets must be properly torqued to ensure correct sensor operation before an air bag system is powered up. If sensor fasteners do not have the proper torque, improper air bag deployment may result in injury to vehicle occupants.

When working on the electrical system on an air-bag-equipped vehicle, use only the vehicle manufacturer's recommended tools and service procedures. The use of improper tools or service procedures may cause accidental air bag deployment and personal injury. For example, do not use 12V or self-powered test lights when servicing the electrical system on an air-bag-equipped vehicle.

Tool and Equipment Safety

Careless use of simple hand tools such as wrenches, screwdrivers, and hammers causes many shop accidents that could be prevented. Keep all hand tools grease-free and in good condition. Tools that slip can cause cuts and bruises. If a tool slips and falls into a moving part, it can fly out and cause serious injury.

Use the proper tool for the job. Make sure the tool is of professional quality. Using poorly made tools or the wrong tools can damage parts or the tool itself, or could cause injury. Never use broken or damaged tools.

Safety around power tools is very important. Serious injury can result from carelessness. Always wear safety glasses when using power tools. If the tool is electrically powered, make sure it is properly grounded. Before using it, check the wiring for cracks in the insulation, as well as for bare wires. Also, when using electrical power tools, never stand on a wet or damp floor. Never leave a running power tool unattended.

When using compressed air, safety glasses and/or a face shield should be worn. Particles of dirt and pieces of metal, blown by the high-pressure air, can penetrate your skin or get into your eyes.

Always be careful when raising a vehicle on a lift or a hoist. Adapters and hoist plates must be positioned correctly to prevent damage to the underbody of the vehicle. There are specific lift points that allow the weight of the vehicle to be evenly supported by the adapters or hoist plates. The correct lift points can be found in the vehicle's service manual. Before operating any lift or hoist, carefully read the operating manual and follow the operating instructions.

Once you feel the lift supports are properly positioned under the vehicle, raise the lift until the supports contact the vehicle. Then, check the supports to make sure they are in full contact with the vehicle. Shake the vehicle to make sure it is securely balanced on the lift, and then raise the lift to the desired working height. Before working under a car, make sure the lift's locking devices are engaged.

A vehicle can be raised off the ground by a hydraulic jack. The jack's lifting pad must be positioned under an area of the vehicle's frame or at one of the manufacturer's recommended lift points. Never place the pad under the floor pan or under steering and suspension components, because these are easily damaged by the weight of the vehicle. Always position the jack so the wheels of the vehicle can roll as the vehicle is being raised.

Safety stands, also called jack stands, should be placed under a sturdy chassis member, such as the frame or axle housing, to support the vehicle after it has been raised by a jack. Once the safety stands are in position, the hydraulic pressure in the jack should be slowly released until the weight of the vehicle is on the stands. Never move under a vehicle when it is supported only by a hydraulic jack. Rest the vehicle on the safety stands before moving under the vehicle.

Parts cleaning is a necessary step in most repair procedures. Always wear the appropriate protection when using chemical, abrasive, and thermal cleaners.

Vehicle Operation

When the customer brings a vehicle in for service, certain driving rules should be followed to ensure your safety and the safety of those working around you. For example, before moving a car into the shop, buckle your safety belt. Make sure no one is near, the way is clear, and there are no tools or parts under the car before you start the engine. Check the brakes before putting the vehicle in gear. Then, drive slowly and carefully in and around the shop.

If the engine must be running while you are working on the car, block the wheels to prevent the car from moving. Place the transmission into park for automatic transmissions or into neutral for manual transmissions. Set the parking (emergency) brake. Never stand directly in front of or behind a running vehicle.

Run the engine only in a well-ventilated area to avoid the danger of poisonous carbon monoxide (CO) in the engine exhaust. CO is an odorless but deadly gas. Most shops have an exhaust ventilation system, and you should always use it. Connect the hose from the vehicle's tailpipe to the intake for the vent system. Make sure the vent system is turned on before running the engine. If the work area does not have an exhaust venting system, use a hose to direct the exhaust out of the building.

HAZARDOUS MATERIALS AND WASTES

A typical shop contains many potential health hazards for those working in it. These hazards can cause injury, sickness, health impairments, discomfort, and even death. Here is a short list of the different classes of hazards:

- Chemical hazards are caused by high concentrations of vapors, gases, or solids in the form of dust.
- Hazardous wastes are those substances that result from a service being performed.
- Physical hazards include excessive noise, vibration, pressure, and temperature.
- Ergonomic hazards are conditions that impede normal and/or proper body position and motion.

There are many government agencies charged with ensuring safe work environments for all workers. These include the Occupational Safety and Health Administration (OSHA), Mine Safety and Health Administration (MSHA), and National Institute for Occupational Safety and Health (NIOSH). These, as well as state and local governments, have instituted regulations that must be understood and followed. Everyone in a shop has the responsibility for adhering to these regulations.

An important part of a safe work environment is the employees' knowledge of potential hazards. Right-to-know laws concerning all chemicals protect every employee in the shop. The general intent of right-to-know laws is to ensure that employers provide their employees with a safe working place as far as hazardous materials are concerned.

All employees must be trained about their rights under the legislation, the nature of the hazardous chemicals in their workplace, and the contents of the labels on the chemicals. All of the information about each chemical must be posted on material safety data sheets (MSDS) and must be accessible. The manufacturer of the chemical must give these sheets to its customers, if they are requested to do so. The sheets detail the chemical composition and precautionary information for all products that can present a health or safety hazard.

Employees must become familiar with the general uses, protective equipment, accident or spill procedures, and any other information regarding the safe handling of a particular hazardous material. This training must be given to employees annually and provided to new employees as part of their job orientation.

All hazardous material must be properly labeled, indicating what health, fire, or reactivity hazard it poses and what protective equipment is necessary when handling each chemical. The manufacturer of the hazardous materials must provide all warnings and precautionary information, which must be read and understood by the user before use. A list of all hazardous materials used in the shop must be posted for the employees to see.

Shops must maintain documentation on the hazardous chemicals in the workplace, proof of training programs, records of accidents or spill incidents, satisfaction of employee requests for specific chemical information via the MSDS, and a general right-to-know compliance procedure manual utilized within the shop.

When handling any hazardous materials or hazardous waste, make sure you follow the required procedures for handling such material. Also wear the proper safety equipment listed on the MSDS. This includes the use of approved respirator equipment.

Some of the common hazardous materials that automotive technicians use are: cleaning chemicals, fuels (gasoline and diesel), paints and thinners, battery electrolyte (acid), used engine oil, refrigerants, and engine coolant (antifreeze).

Many repair and service procedures generate what are known as hazardous wastes. Dirty solvents and cleaners are good examples of hazardous wastes. Something is classified as a hazardous waste if it is on the EPA list of known harmful materials or has one or more of the following characteristics.

- *Ignitability.* If it is a liquid with a flash point below 140°F or a solid that can spontaneously ignite.
- *Corrosivity.* If it dissolves metals and other materials or burns the skin.
- *Reactivity.* Any material that reacts violently with water or other materials or releases cyanide gas, hydrogen sulfide gas, or similar gases when exposed to low pH acid solutions. This also includes material that generates toxic mists, fumes, vapors, and flammable gases.
- *Toxicity.* Materials that leach one or more of eight heavy metals in concentrations greater than 100 times primary drinking water standard concentrations.

Complete EPA lists of hazardous wastes can be found in the Code of Federal Regulations. It should be noted that no material is considered hazardous waste until the shop is finished using it and ready to dispose of it.

The following list covers the recommended procedure for dealing with some of the common hazardous wastes. Always follow these and any other mandated procedures.

Oil Recycle oil. Set up equipment, such as a drip table or screen table with a used oil collection bucket, to collect oils dripping off parts. Place drip pans underneath vehicles that are leaking fluids onto the storage area. Do not mix other wastes with used oil, except as allowed by your recycler. Used oil generated by a shop (and/or oil received from household "do-it-yourself" generators) may be burned on site in a commercial space heater. Also, used oil may be burned for energy recovery. Contact state and local authorities to determine requirements and to obtain the necessary permits.

Oil filters Drain for at least 24 hours, crush, and recycle used oil filters.

Batteries Recycle batteries by sending them to a reclaimer or back to the distributor. Keeping shipping receipts can demonstrate that you have done the recycling. Store batteries in a watertight, acid-resistant container. Inspect batteries for cracks and leaks when they come in. Treat a dropped battery as if it were cracked. Acid residue is hazardous because it is corrosive and may contain lead and other toxic substances. Neutralize spilled acid, by using baking soda or lime, and dispose of it as hazardous material.

Metal residue from machining Collect metal filings when machining metal parts. Keep them separate and recycle if possible. Prevent metal filings from falling into a storm sewer drain.

Refrigerants Recover and/or recycle refrigerants during the servicing and disposal of motor vehicle air conditioners and refrigeration equipment. It is not allowable to knowingly vent refrigerants to the atmosphere.

Recovering and/or recycling during servicing must be performed by an EPA-certified technician using certified equipment and following specified procedures.

Solvents Replace hazardous chemicals with less toxic alternatives that perform equally. For example, substitute water-based cleaning solvents for petroleum-based solvent degreasers. To reduce the amount of solvent used when cleaning parts, use a two-stage process: dirty solvent followed by fresh solvent. Hire a hazardous waste management service to clean and recycle solvents. (Some spent solvents must be disposed of as hazardous waste, unless recycled properly). Store solvents in closed containers to prevent evaporation. Evaporation of solvents contributes to ozone depletion and smog formation. In addition, the residue from evaporation must be treated as a hazardous waste. Properly label spent solvents and store them on drip pans or in diked areas and only with compatible materials.

Containers Cap, label, cover, and properly store, aboveground outdoor liquid containers and small tanks within a diked area and on a paved impermeable surface to prevent spills from running into surface or ground water.

Other solids Store materials such as scrap metal, old machine parts, and worn tires under a roof or tarpaulin to protect them from the elements and to prevent the possibility of creating contaminated runoff. Consider recycling tires by retreading them.

Liquid recycling Collect and recycle coolants from radiators. Store transmission fluids, brake fluids, and solvents containing chlorinated hydrocarbons separately, and recycle or dispose of them properly.

Shop towels and rags Keep waste towels in a closed container marked "contaminated shop towels only." To reduce costs and liabilities associated with disposal of used towels, which can be classified as hazardous wastes, investigate using a laundry service that is able to treat the wastewater generated from cleaning the towels.

Waste storage Always keep hazardous waste separate, properly labeled, and sealed in the recommended containers. The storage area should be covered and may need to be fenced and locked if vandalism could be a problem. Select a licensed hazardous waste hauler after seeking recommendations and reviewing the firm's permits and authorizations.

ENGINE PERFORMANCE TOOLS AND EQUIPMENT

Many different tools and many kinds of testing and measuring equipment are used to service engine performance systems. NATEF has identified many of these and has said an Engine Performance technician must know what they are and how and when to use them. The tools and equipment listed by NATEF are covered in the following discussion. Also included are the tools and equipment you will use while completing the job sheets. Although you need to be more than familiar with and will be using common hand tools, they are not part of this discussion. You should already know what they are and how to use and care for them.

Scan Tools

The introduction of computer-controlled systems brought with it the need for tools capable of troubleshooting electronic control systems. There are a variety of computer scan tools available today that do just that. A scan tool is a microprocessor designed to communicate with the vehicle's computer. Connected to the computer through diagnostic connectors, a scan tool can access trouble codes, run tests to check system operations, and monitor the activity of the system (Figure 2).

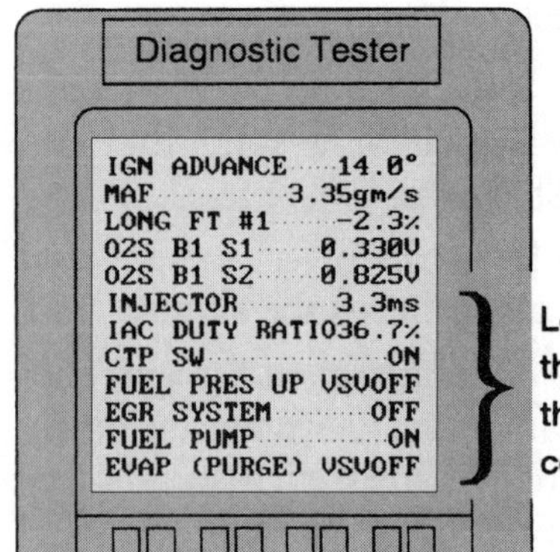

Figure 2 A sample of the information that can be retrieved on a scan tool.

Trouble codes and test results are displayed on an LED screen, or printed out on the scanner printer.

Scan tools are capable of testing many on-board computer systems, such as climate controls, transmission controls, engine computers, antilock brake computers, air bag computers, and suspension computers, depending on the year and make of the vehicle and the type of scan tester. In many cases, the technician must select the computer system to be tested with the scanner after it has been connected to the vehicle.

The scan tool is connected to specific diagnostic connectors on the vehicle. Some manufacturers have one diagnostic connector. This connects the data wire from each computer to a specific terminal in this connector. Other manufacturers have several diagnostic connectors on each vehicle, and each of these connectors may be connected to one or more computers. The scan tool must be programmed for the model year, make of vehicle, and type of engine.

With OBD-II, the diagnostic connectors are located in the same place on all vehicles. Also, any scan tool designed for OBD-II will work on all OBD-II systems, therefore the need to have designated scan tools or cartridges is eliminated. Most OBD-II scan tools have the ability to store, or "freeze" data during a road test, and then play back this data when the vehicle is returned to the shop.

There are many different scan tools available. Some are a combination of other diagnostic tools, such as a lab scope and graphing multimeter. These may have the following capabilities:

- Retrieve DTCs
- Monitor system operational data
- Reprogram the vehicle's electronic control modules
- Perform systems diagnostic tests
- Display appropriate service information, including electrical diagrams
- Display TSBs
- Troubleshooting instructions
- Easy tool updating through a personal computer (PC)

The vehicle's computer sets trouble codes when a voltage signal is entirely out of its normal range. The codes help technicians identify the cause of the problem when this is the case. If a signal is within its normal range but is still not correct, the vehicle's computer will not display a trouble code. However, a problem may still exist.

Some scan tools work directly with a PC through uncabled communication links, such as Bluetooth. Others use a Personal computer or even a cell phone with special programming. These are small hand-held units that allow you to read diagnostic trouble codes (DTCs), monitor the activity of sensors, and inspection/maintenance system test results to quickly determine what service the vehicle requires. Most of these scan tools also have the ability to:

- Perform system and component tests
- Report test results of monitored systems
- Exchange files between a PC and PDA
- View and print files on a PC
- Print DTC/Freeze Frame
- Generate emissions reports
- IM/Mode 6 information
- Display related TSBs
- Display full diagnostic code descriptions
- Observe live sensor data
- Update the scan tool as a manufacturer's interfaces change

Lab Scopes

An oscilloscope is a visual voltmeter. An oscilloscope converts electrical signals to a visual image representing voltage changes over a specific period of time. This information is displayed in the form of a continuous voltage line called a waveform pattern or trace.

An upward movement of the voltage trace on an oscilloscope screen indicates an increase in voltage, and a downward movement of this trace represents a decrease in voltage. As the voltage trace moves across an oscilloscope screen, it represents a specific length of time.

The size and clarity of the displayed waveform is dependent on the voltage scale and the time reference selected. Most scopes are equipped with controls that allow voltage and time interval selection. It is important, when choosing the scales, to remember that a scope displays voltage over time.

Dual-trace oscilloscopes can display two different waveform patterns at the same time. This

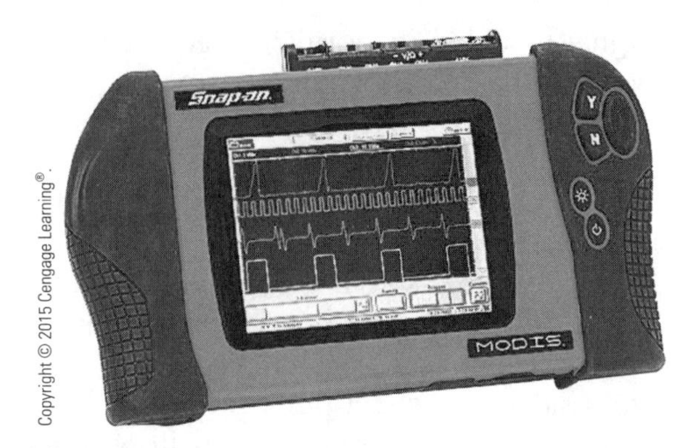

Figure 3 A DSO with multiple trace capability.

makes cause and effect analysis easier. Other scopes are capable of displaying more than two traces (Figure 3).

With a scope, precise measurement is possible. A scope will display any change in voltage as it occurs. This is especially important for diagnosing intermittent problems. It is also invaluable for checking the primary and secondary circuits, as well as fuel injectors and various inputs and outputs of the control system.

The screen of a lab scope is divided into small divisions of time and voltage. Time is represented by the horizontal movement of the waveform. Voltage is measured with the vertical position of the waveform. Since the scope displays voltage over time, the waveform moves from the left (the beginning of measured time) to the right (the end of measured time). The value of the divisions can be adjusted to improve the view of the voltage waveform.

Since a scope displays actual voltage, it will display any electrical noise or disturbances that accompany the voltage signal. Noise is primarily caused by radio frequency interference (RFI), which may come from the ignition system. RFI is an unwanted voltage signal that rides on a signal. This noise can cause intermittent problems with unpredictable results. The noise causes slight increases and decreases in the voltage. When a computer receives a voltage signal with noise, it will try to react to the minute changes. As a result, the computer responds to the noise rather than the voltage signal.

Graphing Multimeter

One of the best investments in diagnostic tools is a graphing digital multimeter. These meters display readings over time, similar to a lab scope. The graph displays the minimum and maximum readings on a graph, as well as displaying the current reading. By observing the graph, a technician can observe any undesirable changes during the transition from a low reading to a high reading, or vice versa. These glitches are some of the more difficult problems to identify without a graphing meter or a lab scope.

Exhaust Analyzers

Federal laws require that new cars and light trucks must meet specific emissions levels. State governments have also passed laws requiring that car owners maintain their vehicles so that the emissions remain below an acceptable level. Most states require an annual emissions inspection to meet that goal. Many shops have an exhaust analyzer for inspection purposes.

Exhaust analyzers (Figure 4) are also very valuable diagnostic tools. By looking at the quality of an engine's exhaust, a technician is able to observe the effects of the combustion process. Any defect can cause a change in exhaust quality. The amount and type of change serves as the basis of diagnostic work.

The manufacturers of exhaust analyzers have altered their machines so that they can look at the efficiency of an engine, in spite of the effectiveness of the emission controls. These machines are four-gas exhaust analyzers. In addition

Figure 4 An exhaust gas analyzer.

to measuring HC and CO levels, a four-gas exhaust analyzer also monitors carbon dioxide (CO_2) and oxygen (O_2) levels in the exhaust. The latter two gases are changed only slightly by the emission controls and therefore can be used to check engine efficiency. Many exhaust analyzers are also available that measure a fifth gas, oxides of nitrogen (NOx).

By measuring oxides of nitrogen (NOx), carbon dioxide (CO_2), and oxygen (O_2), in addition to HC and CO, a technician gets a better look at the efficiency of the engine. Keep in mind that an exhaust analyzer is an excellent diagnostic tool and is not used just for comparing emissions levels against standards. There is a desired relationship between the five gases. Any deviation from this relationship can be used to diagnose a driveability problem.

Ideally, the combustion process will combine fuel (HC) and oxygen (O_2) to form water and carbon dioxide (CO_2). Although most of the air brought into the engine is nitrogen, this gas should not become part of the combustion process and should pass out of the exhaust as nitrogen. NOx is formed only when there are very high combustion temperatures. High amounts of NOx can be caused by cooling system problems, carbon build-up on the pistons (causing higher than normal compression ratios), a defective EGR system, and lean mixtures.

Small amounts of CO_2 are normally present in our air and CO_2 is not considered a pollutant. CO_2 is a desired element in the exhaust stream and is only present when there is complete combustion. Therefore, the more CO_2 in the exhaust stream, the better.

O_2 is used to oxidize CO and HC into water and CO_2. The O_2 level is an indicator of air/fuel mixture. As the mixture goes lean, O_2 levels increase. As the mixture goes rich, O_2 moves and stays low. Ideally, we would like to see very low amounts of O_2 in the exhaust stream.

Vacuum Gauge

Measuring intake manifold vacuum is another way to diagnose the condition of an engine. Manifold vacuum is tested with a vacuum gauge. Vacuum is formed on a piston's intake stroke. As the piston moves down, it lowers the pressure of the air in the cylinder—if the cylinder is sealed. This lower cylinder pressure is called engine vacuum. If there is a leak, atmospheric pressure will force air into the cylinder and the resultant pressure will not be as low. The reason atmospheric pressure enters is simply that whenever there is a low and high pressure, the high pressure will always move toward the low pressure. Vacuum is measured in inches of mercury (in./Hg) and in kilopascals (kPa) or millimeters of mercury (mm/Hg).

To measure vacuum, a flexible hose on the vacuum gauge is connected to a source of manifold vacuum, either on the manifold or a point below the throttle plates. The test is made with the engine cranking or running. A good vacuum reading is typically at least 16 in./Hg. However, a reading of 15 to 20 in./Hg (50 to 65 kPa) is normally acceptable. Since the intake stroke of each cylinder occurs at a different time, the production of vacuum occurs in pulses. If the amount of vacuum produced by each cylinder is the same, the vacuum gauge will show a steady reading. If one or more cylinders are producing different amounts of vacuum, the gauge will show a fluctuating reading.

Circuit Tester

Circuit testers are used to identify short and open circuits in any electrical circuit. Low-voltage testers are used to troubleshoot 6- to 12-volt circuits. A circuit tester, commonly called a test light, looks like a stubby ice pick. Its handle is transparent and contains a light bulb. A probe extends from one end of the handle and a ground clip and wire from the other end. When the ground clip is attached to a good ground and the probe touched to a live connector, the bulb in the handle will light up. If the bulb does not light, voltage is not available at the connector.

WARNING: *Do not use a conventional 12-volt test light to diagnose components and wires in electronic systems. The current draw of these test lights may damage computers and system components. High-impedance test lights are available for diagnosing electronic systems.*

A self-powered test light is called a continuity tester. It is used on non-powered circuits. It looks like a regular test light, except that it has a small internal battery. When the ground clip is attached to the negative side of a component and the probe touched to the positive side, the lamp will light if there is continuity in the circuit. If an open circuit exists, the lamp will not light. Do not use any type of test light or circuit tester to diagnose automotive air bag systems.

Voltmeter

A voltmeter has two leads: a red positive lead and a black negative lead. The red lead should be connected to the positive side of the circuit or component. The black should be connected to ground or to the negative side of the component. Voltmeters should be connected across the circuit being tested.

The voltmeter measures the voltage available at any point in an electrical system. A voltmeter can also be used to test voltage drop across an electrical circuit, component, switch, or connector. A voltmeter can also be used to check for proper circuit grounding.

Ohmmeter

An ohmmeter measures the resistance to current flow in a circuit. In contrast to the voltmeter, which uses the voltage available in the circuit, the ohmmeter is battery powered. The circuit being tested must be open. If the power is on in the circuit, the ohmmeter will be damaged.

The two leads of the ohmmeter are placed across or in parallel with the circuit or component being tested. The red lead is placed on the positive side of the circuit and the black lead is placed on the negative side of the circuit. The meter sends current through the component and determines the amount of resistance based on the voltage dropped across the load. The scale of an ohmmeter reads from zero to infinity. A zero reading means there is no resistance in the circuit and may indicate a short in a component that should show a specific resistance. An infinite reading indicates a number higher than the meter can measure. This usually is an indication of an open circuit.

Ohmmeters are also used to trace and check wires or cables. Assume that one wire of a four-wire cable is to be found. Connect one probe of the ohmmeter to the known wire at one end of the cable and touch the other probe to each wire at the other end of the cable. Any evidence of resistance, such as meter needle deflection, indicates the correct wire. Using this same method, you can check a suspected defective wire. If resistance is shown on the meter, the wire is sound. If no resistance is measured, the wire is defective (open). If the wire is okay, continue checking by connecting the probe to other leads. Any indication of resistance indicates that the wire is shorted to one of the other wires and that the harness is defective.

Ammeter

An ammeter measures current flow in a circuit. The ammeter must be placed into the circuit or in series with the circuit being tested. Normally, this requires disconnecting a wire or connector from a component and connecting the ammeter between the wire or connector and the component. The red lead of the ammeter should always be connected to the side of the connector closest to the positive side of the battery and the black lead should be connected to the other side.

It is much easier to test current using an ammeter with an inductive pickup. The pickup clamps around the wire or cable being tested. These ammeters measure amperage based on the magnetic field created by the current flowing through the wire. This type of pickup eliminates the need to separate the circuit to insert the meter.

Because ammeters are built with very low internal resistance, connecting them in series does not add any appreciable resistance to the circuit. Therefore, an accurate measurement of the current flow can be taken.

Volt/Ampere Tester

A volt/ampere tester (VAT) is used to test batteries, starting systems, and charging systems. The tester contains a voltmeter, ammeter, and carbon pile. The carbon pile is a variable resistor. A knob on the tester allows the technician to vary the resistance of the pile. When the tester is attached to the battery, the carbon pile will draw current out of the battery. The ammeter will read the amount of current draw. When testing a battery, the resistance of the carbon pile is adjusted so the current draw matches the ratings of the battery.

Logic Probes

In some circuits pulsed or digital signals pass through the wires. These "on-off" digital signals either carry information or provide power to drive a component. Many sensors used in a computer-control circuit send digital information back to the computer. To check the continuity of the wires that carry digital signals, a logic probe can be used.

A logic probe has three different colored LEDs. A red LED lights when there is high voltage at the point being probed. A green LED lights to indicate low voltage. And a yellow LED indicates the presence of a voltage pulse. The logic probe is powered by the circuit and reflects only the activity at the point being probed. When the probe's test leads are attached to a circuit, the LEDs display the activity.

If a digital signal is present, the yellow LED will turn on. When there is no signal, the LED is off. If voltage is present, the red or green LEDs will light, depending on the amount of voltage. When there is a digital signal and the voltage cycles from low to high, the yellow LED will be lit and the red and green LEDs will cycle indicating a change in the voltage.

DMMs

It's not necessary for a technician to own separate meters to measure volts, ohms, and amps, because a multimeter can be used instead. Top-of-the-line multimeters are multifunctional. Most test volts, ohms, and amperes in both DC and AC. Usually there are several test ranges provided for each of these functions. In addition to these basic electrical tests, multimeters also test engine rpm, duty cycle, pulse width, diode condition, frequency, and even temperature. The technician selects the desired test range by turning a control knob on the front of the meter.

Multimeters are available with either analog or digital displays. But, the most commonly used multimeter is the digital volt/ohmmeter (DVOM), which is often referred to as a digital multimeter (DMM). There are several drawbacks to using analog-type meters for testing electronic control systems. Many electronic components require very precise test results. Digital meters can measure volts, ohms, or amperes in tenths and hundredths. Another problem with analog meters is their low internal resistance (input impedance). The low input impedance allows too much current to flow through circuits and should not be used on delicate electronic devices.

Digital meters, on the other hand, have high input impedance, usually at least 10 megohms (10 million ohms). Metered voltage for resistance tests is well below 5 volts, reducing the risk of damage to sensitive components and delicate computer circuits. A high-impedance digital multimeter must be used to test the voltage of some components and systems such as an oxygen (O_2) sensor circuit. If a low-impedance analog meter is used in this type of circuit, the current flow through the meter is high enough to damage the sensor.

DMMs either have an "auto range" feature, in which the meter automatically selects the appropriate scale, or they must be set to a particular range. In either case, you should be familiar with the ranges and the different settings available on the meter you are using. To designate particular ranges and readings, meters display a prefix before the reading or range. If the meter has a setting for mAmps, this means the readings will be given in milli-amps or 1/1000th of an amp. Ohmmeter scales are expressed as a multiple of tens or use the prefix K or M. K stands for Kilo or 1000. A reading of 10K ohms equals 10,000 ohms. An M stands for Mega or 1,000,000. A reading of 10M ohms equals 10,000,000 ohms. When using a meter with an auto range, make sure you note the range being used by the meter. There is a big difference between 10 ohms and 10,000,000 ohms.

WARNING: *Many digital multimeters with auto range display the measurement with a decimal point.*

After the test range has been selected, the meter is connected to the circuit in the same way as if it were an individual meter.

When using the ohmmeter function, the DMM will show a zero or close to zero when there is good continuity. If the continuity is very poor, the meter will display an infinite reading. This reading is usually shown as a blinking "1.000," a blinking "1," or an "OL." Before taking any measurement, calibrate the meter. This is done by holding the two leads together and adjusting the meter reading to zero. Not all meters need to be calibrated; some digital meters automatically calibrate when a scale is selected. On meters that require calibration, it is recommended that the meter be zeroed after changing scales.

Multimeters may also have the ability to measure duty cycle, pulse width, and frequency. All of these represent voltage pulses caused by the turning on and off of a circuit or the increase and decrease of voltage in a circuit. Duty cycle is a measurement of the amount of time something is on compared to the time of one cycle and is measured in a percentage.

Pulse width is similar to duty cycle except that it is the exact time something is turned on and is measured in milliseconds. When measuring duty cycle, you are looking at the amount of time something is on during one cycle.

The number of cycles that occur in one second is called the frequency. The higher the frequency, the more cycles occur in a second. Frequencies are measured in Hertz. One Hertz is equal to one cycle per second.

Hybrid Tools

A hybrid vehicle is an automobile and as such is subject to many of the same problems as a conventional vehicle. Most systems in a hybrid vehicle are diagnosed in the same way as well. However, a hybrid vehicle has unique systems that require special procedures and test equipment. It is imperative to have good information before attempting to diagnose these vehicles. Also, make sure you follow all test procedures precisely as they are given.

An important diagnostic tool is a DMM. However, this is not the same DMM used on a conventional vehicle. The meter used on hybrids and other electric-powered vehicles should be classified as a category III meter. There are basically four categories for low-voltage electrical meters, each built for specific purposes and to meet certain standards. Low-voltage, in this case, means voltages less than 1000-volts. The categories define how safe a meter is when measuring certain circuits. The standards for the various categories are defined by the American National Standards Institute (ANSI), the International Electrotechnical Commission (IEC), and the Canadian Standards Association (CSA). A CAT III meter is required for testing hybrid vehicles because of the high-voltages, three-phase current, and the potential for high transient voltages. Transient voltages are voltage surges or spikes that occur in AC circuits. To be safe, you should have a CAT III-1000 V meter. A meter's voltage rating reflects its ability to withstand transient voltages. Therefore, a CAT III 1000 V meter offers much more protection than a CAT III meter rated at 600 volts.

Another important tool is an insulation resistance tester. These can check for voltage leakage through the insulation of the high-voltage cables. Obviously no leakage is desired and any leakage can cause a safety hazard as well as damage to the vehicle. Minor leakage can also cause hybrid system-related driveability problems. This meter is not one commonly used by automotive technicians, but should be for anyone who might service a damaged hybrid vehicle, such as doing body repair. This should also be a CAT III meter and may be capable of checking resistance and voltage of circuits like a DMM.

To measure insulation resistance, system voltage is selected at the meter and the probes placed at their test position. The meter will display the voltage it detects. Normally, resistance readings are taken with the circuit de-energized unless you are checking the effectiveness of the cable or wire insulation. In this case, the meter is measuring the insulation's effectiveness and not its resistance.

The probes for the meters should have safety ridges or finger positioners. These help prevent physical contact between your fingertips and the meter's test leads.

Static Safeguards

Some manufacturers mark certain components and circuits with a code or symbol to warn technicians that they are sensitive to electrostatic discharge. Static electricity can destroy or render a component useless.

When handling any electronic part, especially those that are static sensitive, follow the guidelines below to reduce the possibility of electrostatic build-up on your body and the inadvertent discharge to the electronic part. If you are not sure if a part is sensitive to static, treat it as if it were.

1. Always touch a known good ground before handling the part. This should be repeated while handling the part and more frequently after sliding across a seat, sitting down from a standing position, or walking a distance.
2. Avoid touching the electrical terminals of the part, unless you are instructed to do so in the written service procedures. It is good practice to keep your fingers off all electrical terminals, as the oil from your skin can cause corrosion.
3. When you are using a voltmeter, always connect the negative meter lead first.

4. Do not remove a part from its protective package until it is time to install the part.
5. Before removing the part from its package, ground yourself and the package to a known good ground on the vehicle.

Some tool manufacturers have grounding straps that are available. These are designed to fit on you, at one end, and be fastened to a good ground on the other end. They are some of the best safeguards against static electricity.

Computer Memory Saver

Memory savers are an external power source used to maintain the memory circuits in electronic accessories and the engine, transmission, and body computers when the vehicle's battery is disconnected. The saver is plugged into the vehicle's cigar lighter outlet. It can be powered by a 9- or 12-volt battery.

Feeler Gauge

A feeler gauge is a thin strip of metal or plastic of known and closely controlled thickness. Several of these strips are often assembled together as a feeler gauge set that looks like a pocketknife. The desired thickness gauge can be pivoted away from the others for convenient use. A feeler gauge set usually contains strips or leaves of 0.002- to 0.010-inch thickness (in steps of 0.001 inch) and leaves of 0.012- to 0.024-inch thickness (in steps of 0.002 inch).

A feeler gauge can be used by itself to measure clearances and end play. Round, wire feeler gauges are often used to measure spark plug gap. The round gauges are designed to give a better feel for the fit of the gauge in the gap.

A specially designed tool measures and corrects the gap. This pliers-like tool utilizes flat gauges for the adjustment procedure. The gauges are mounted on the tool like spokes on a wheel. Above the gauges is an anvil, which is used to apply pressure to the electrode. On the opposite end of the tool is a curved seat. This seat performs two functions: it supports the plug shell during the procedure and also compresses the ground electrode against the gauge, thus setting the air gap.

A tapered regapping tool is simply a piece of tapered steel with leading and trailing edges of different dimensions. Between these two points, the gauge varies in thickness. A scale, located above the gauge, indicates the thickness at any given point. When the gauge is slid between electrodes, it stops when the air gap size reaches the thickness on the gauge. The scale reading is made in thousandths of an inch. Adjusting slots are available to bend the ground electrode as needed to make the air gap adjustment.

Torque Wrench

Torque is the twisting force used to turn a fastener against the friction between the threads and between the head of the fastener and the surface of the component. The fact that practically every vehicle and engine manufacturer publishes a list of torque recommendations is ample proof of the importance of using proper amounts of torque when tightening nuts or bolts. The amount of torque applied to a fastener is measured with a torque-indicating or torque wrench.

A torque wrench is basically a ratchet or breaker bar with some means of displaying the amount of torque exerted on a bolt when pressure is applied to the handle. Torque wrenches are available with the various drive sizes. Sockets are inserted onto the drive and then placed over the bolt. As pressure is exerted on the bolt, the torque wrench indicates the amount of torque.

The common types of torque wrenches are available with inch-pound and foot-pound increments.

- A beam torque wrench is not highly accurate. It relies on a beam metal that points to the torque reading.
- A "click"-type torque wrench clicks when the desired torque is reached. The handle is twisted to set the desired torque reading.
- A dial torque wrench has a dial that indicates the torque exerted on the wrench. The wrench may have a light or buzzer that turns on when the desired torque is reached.
- A digital readout type displays the torque and is commonly used to measure turning effort, as well as for tightening bolts. Some designs of this type torque wrench have a light or buzzer that turns on when the desired torque is reached.

An important item that many forget to torque properly is the spark plug. A spark plug is a plug that allows a spark to enter the combustion chamber without allowing combustion gases to leak. The latter purpose is the reason spark plugs need to be properly tightened in their bores.

Spark Plug Thread Repair

Sometimes when spark plugs are removed from a cylinder head, the threads have traces of metal on them. This happens more often with aluminum heads. When this occurs, the spark plug bore must be corrected by installing thread inserts.

When installing spark plugs, if the plugs cannot be installed easily by hand, the threads in the cylinder head may need to be cleaned with a thread-chasing tap. Be especially careful not to cross-thread the plugs when working with aluminum heads. Always tighten the plugs with a torque wrench and the correct spark plug socket, following the vehicle manufacturer's specifications.

Spark Tester

A spark tester is a fake spark plug. The tester is constructed like a spark plug but doesn't have a ground electrode. In place of the electrode there is a grounding clamp. Using test spark plugs is an easy way to determine if the ignition problem is caused by something in the primary or secondary circuit.

The spark tester is inserted in the spark plug end of an ignition cable. When the engine is cranked, a spark should be seen from the tester to a ground. Experience with these testers will also help you determine the intensity of the spark.

Sensor Tools

Because they are shaped much like a spark plug with wires or a connector coming out of the top, ordinary sockets don't fit well on an oxygen sensor. For this reason, tool manufacturers provide special sockets for these sensors.

Special sockets are also available for other sending units and sensors.

Fuel Pressure Gauge

A fuel pressure gauge (Figure 5) is very important for diagnosing fuel injection systems. These systems rely on high fuel pressures, from 35 to 70 psi. A drop in fuel pressure will reduce the amount of fuel delivered to the injectors and result in a lean air/fuel mixture.

Figure 5 A fuel pressure gauge.

A fuel pressure gauge is used to check the discharge pressure of fuel pumps, the regulated pressure of fuel injection systems, and injector pressure drop. This test can identify faulty pumps, regulators, or injectors and can identify restrictions present in the fuel delivery system. Restrictions are typically caused by a dirty fuel filter, collapsed hoses, or damaged fuel lines.

Some fuel pressure gauges also have a valve and outlet hose for testing fuel pump discharge volume. The manufacturer's specification for discharge volume will be given as a number of pints or liters of fuel that should be delivered in a certain number of seconds.

Injector Balance Tester

The injector balance tester (Figure 6) is used to test the injectors in a port fuel injected engine for proper operation. A fuel pressure gauge is also used during the injector balance test. The injector balance tester contains a timing circuit, and some injector balance testers have an off-on switch. A pair of leads on the tester must be connected to the battery with the correct polarity. The injector terminals are disconnected, and a second double lead on the tester is attached to the injector terminals.

Figure 6 A fuel injector balance tester.

Before conducting an injector test, a fuel pressure gauge is connected to the Schrader valve on the fuel rail, and the ignition switch should be cycled two or three times until the specified fuel pressure is indicated on the pressure gauge. When the tester's push button is depressed, the tester energizes the injector winding for a specific length of time, and the technician records the pressure decrease on the fuel pressure gauge. This procedure is repeated on each injector.

Some vehicle manufacturers provide a specification of 3-psi (20 kPa) maximum difference between the pressure readings after each injector is energized. If the injector orifice is restricted, there is not much pressure decrease when the injector is energized. Acceleration stumbles, engine stalling, and erratic idle operation are caused by restricted injector orifices. The injector plunger is sticking open if excessive pressure drop occurs when the injector is energized. Sticking injector plungers may result in a rich air/fuel mixture.

Because electronic fuel injection systems are pressurized, make sure you depressurize the system before opening up the system for diagnostics or repair work.

Injector Circuit Test Light

A special test light called a "noid light" can be used to determine if a fuel injector is receiving its proper voltage pulse from the computer. The wiring harness connector is disconnected from the injector and the noid light is plugged into the connector. After disabling the ignition to prevent starting, the engine is turned over by the starter motor. The noid light will flash rapidly if the voltage signal is present. No flash usually indicates an open in the power feed or ground circuit to the injector.

Fuel Injector Cleaners

Fuel injectors spray a certain amount of fuel into the intake system. If the fuel pressure is low, not enough fuel will be sprayed. This is also true if the fuel injector is dirty. Normally clogged injectors are the result of inconsistencies in gasoline detergent levels and the high sulfur content of gasoline. When these sensitive fuel injectors become partially clogged, fuel flow is restricted. Spray patterns are altered, causing poor performance and reduced fuel economy.

The solution to a sulfated and/or plugged fuel injector is to clean it, not replace it. There are two kinds of fuel injector cleaners (Figure 7). One is a pressure tank. A mixture of solvent

Figure 7 A typical fuel injector cleaning machine.

and unleaded gasoline is placed in the tank, following the manufacturer's instructions for mixing, quantity, and safe handling. The vehicle's fuel pump must be disabled and, on some vehicles, the fuel line must be blocked between the pressure regulator and the return line. Then, the hose on the pressure tank is connected to the service port in the fuel system. The in-line valve is then partially opened and the engine is started. It should run at approximately 2,000 rpm for about 10 minutes to clean the injectors thoroughly.

An alternative to the pressure tank is a pressurized canister in which the solvent solution is premixed. Use of the canister-type cleaner is similar to this procedure, but does not require mixing or pumping.

The canister is connected to the injection system's servicing fitting, and the valve on the canister is opened. The engine is started and allowed to run until it dies. Then the canister is discarded.

Fuel Line Tools

Many vehicles are equipped with quick-connect line couplers (Figure 8). These work well to seal the connection but are almost impossible to disconnect if the correct tools are not used. There is a variety of quick-connect fittings and tools.

Figure 8 A quick-connect coupling tool.

EVAP Leak Detector

A very popular way to identify EVAP leaks is with a smoke machine. Nearly all OEMs recommend this method. The smoke machine vaporizes a specially formulated, highly refined mineral-oil-based fluid. The machine then introduces the resultant smoky vapor into the EVAP system.

Pressurized nitrogen pushes the smoke through the system. The source of a leak is identified by the escaping smoke. As when pressure testing the system, the system must be sealed to get accurate results. This means the canister vent port must be blocked off. On some vehicles the solenoid can be closed through commands input with a scan tool.

Fuel Cap Tester

Most fuel cap testers come with a variety of fuel cap adapters. Always use the one that is appropriate for the vehicle being tested. Tighten the cap to the adapter and connect the cap and adapter to the tester. Turn on the tester. The tester will create a pressure on the cap and monitor the cap's ability to hold the pressure. In most cases, the tester will illuminate lights indicating the cap is good or bad.

Vacuum Pump

There are many vacuum-operated devices and switches on cars. These devices use engine vacuum to cause a mechanical action or to switch something on or off. The tool used to test vacuum-actuated components is the vacuum pump. There are two types of vacuum pumps: an electrical operated pump and a hand-held pump (Figure 9). The hand-held pump is most often used for diagnostics. A hand-held vacuum pump consists of a hand pump, a vacuum gauge, and a length of rubber hose used to attach the pump to the component being tested. Tests with the vacuum pump can usually be performed without removing the component from the car.

When the handles of the pump are squeezed together, a piston inside the pump body draws air out of the component being tested. The partial vacuum created by the pump is registered on the pump's vacuum gauge. While forming a vacuum in a component, watch the action of the component. The vacuum level needed to actuate a given component should be compared to the specifications given in the factory service manual.

Figure 9 A hand-held vacuum pump.

The vacuum pump is also commonly used to locate vacuum leaks. This is done by connecting the vacuum pump to a suspect vacuum hose or component and applying vacuum. If the needle on the vacuum gauge begins to drop after the vacuum is applied, a leak exists somewhere in the system.

Vacuum Leak Detector

A vacuum or compression leak might be revealed by a compression check, a cylinder leak down test, or a manifold vacuum test. However, finding the location of the leak can often be very difficult.

A simple, but time-consuming way to find leaks in a vacuum system is to check each component and vacuum hose with a vacuum pump. Simply apply vacuum to the suspected area and watch the gauge for any loss of vacuum. A good vacuum component will hold the vacuum that is applied to it.

Another method of leak detection is done by using an ultrasonic leak detector. Air rushing through a vacuum leak creates a high-frequency sound, higher than the range of human hearing. An ultrasonic leak detector is designed to hear the frequencies of the leak. When the tool is passed over a leak, the detector responds to the high-frequency sound by emitting a warning beep. Some detectors also have a series of LEDs that light up as the frequencies are received. The closer the detector is moved to the leak, the more LEDs light up or the faster the beeping occurs. This allows the technician to zero in on the leak. An ultrasonic leak detector can sense leaks as small as 1/500 inch and accurately locate the leak to within 1/16 inch.

An ultrasonic leak detector can also be used to detect the source of compression leaks, bearing wear, and electrical arcing. It can also be used to diagnose fuel injector operation. Some technicians also use a smoke tester connected to the intake manifold to check for vacuum leaks.

Pinch-Off Pliers

The need to pinch off a rubber hose is common during diagnostics and service. Special pliers are designed to do this without damaging the hose. These pliers are much like vise-grip pliers in that they hold their position until they are released. The jaws of the pliers are flat and close in a parallel motion. Both of these features prevent damage to the hose.

Pyrometers

The converter should be checked for its ability to convert CO and HC into CO_2 and water. One of the ways to do this is to do a delta temperature test. To conduct this test, use a hand-held digital pyrometer. By touching the pyrometer probe to the exhaust pipe just ahead of and just behind the converter, there should be an increase of at least 100°F or 8% above the inlet temperature reading as the exhaust gases pass through the converter. If the outlet temperature is the same or lower, nothing is happening inside the converter.

Compression Testers

Engines depend on the compression of the air/fuel mixture to have power output. The compression stroke of the piston compresses the air/fuel mixture within the combustion chamber. If the combustion chamber leaks, some of the air/fuel mixture will escape while it is being compressed, resulting in a loss of power and a waste of fuel. The leaks can be caused by burned valves, a blown head gasket, worn rings, a slipped timing belt or chain, worn valve seats, a cracked head, and other factors.

An engine with poor compression (lower compression pressure due to leaks in the cylinder) will not run correctly. If a symptom suggests that the cause of a problem may be poor compression, a compression test is performed.

A compression gauge is used to check cylinder compression. The dial face on the typical compression gauge indicates pressure in both pounds per square inch (psi) and metric kilopascals (kPa). The range is usually 0 to 300 psi and 0 to 2,100 kPa. There are two basic types of compression gauges: a push-in gauge and a screw-in gauge.

The push-in type has a short stem that is either straight or bent at a 45-degree angle. The stem ends in a tapered rubber tip that fits any size spark plug hole. The rubber tip is placed in the spark plug hole, after the spark plugs have been removed, and held there while the engine is cranked through several compression cycles. Although simple to use, the push-in gauge may give inaccurate readings if it is not held tightly in the hole.

The screw-in gauge has a long, flexible hose that ends in a threaded adapter. This type of compression tester is often used because its flexible hose can reach into areas that are difficult to reach with a push-in type tester. The threaded adapters are changeable and come in several thread sizes to fit 10-mm, 12-mm, 14-mm, and 18-mm diameter holes. The adapters screw into the spark plug holes in place of the spark plugs.

Most compression gauges have a vent valve that allows the gauge's meter to hold the highest pressure reading. Opening the valve releases the pressure when the test is complete.

Cylinder Leakage Tester

If a compression test shows that any of the cylinders are leaking, a cylinder leakage test can be performed to measure the percentage of compression lost and help locate the source of leakage.

A cylinder leakage tester (Figure 10) applies compressed air to a cylinder through the spark plug hole. Before the air is applied to the cylinder, the piston of that cylinder must be at TDC on its compression stroke. A threaded adapter on the end of the air pressure hose screws into the spark plug hole. The source of the compressed air is normally the shop's compressed air system. A pressure regulator in the tester controls the pressure applied to the cylinder. An analog gauge registers the percentage of air pressure lost from the cylinder when the compressed air is applied. The scale on the dial face reads 0 to 100 percent.

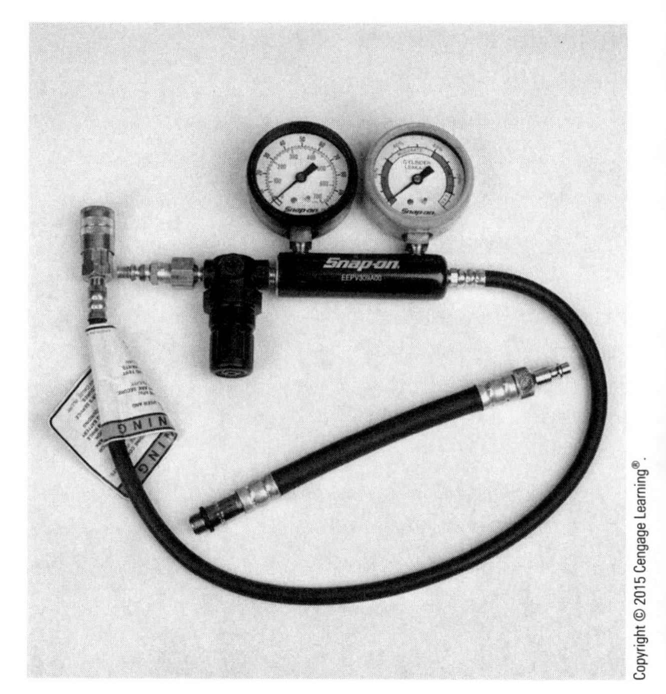

Figure 10 A cylinder leakage tester.

A zero reading means there is no leakage from the cylinder. Readings of 100 percent would indicate that the cylinder does not hold pressure. The location of the compression leak can be found by listening and feeling around various parts of the engine. If air is felt or heard leaving the throttle plate assembly, a leaking intake valve is indicated. If a bad exhaust valve is responsible for the leakage, air can be felt leaving the exhaust system during the test. Air leaving the radiator would indicate a faulty head gasket or a cracked block or head. If the piston rings are bad, air will be heard leaving the valve cover's breather cap or the oil dipstick tube.

Most vehicles, even new cars, experience some leakage around the rings. Up to 20 percent is considered acceptable during the leakage test. When the engine is actually running, the rings will seal much more tightly and the actual percent of leakage will be lower. However, there should be no leakage around the valves or the head gasket.

Cylinder Power Balance Test

The cylinder power balance test is used to see if all of the engine's cylinders are producing the same amount of power. Ideally, all cylinders will produce the same amount. To check an engine's power balance, the spark plugs for individual cylinders are shorted out one at a time and the change in engine speed is recorded. If all of the cylinders are producing the same amount of power, engine speed will drop the same amount as each plug is shorted. Unequal cylinder power balance can mean a problem in the cylinders themselves, as well as the rings, valves, intake manifold, head gasket, fuel system, or ignition system.

A power balance test is performed quickly and easily using an engine analyzer, because the firing of the spark plugs can be automatically controlled or manually controlled by pushing a button. Some vehicles have a power balance test built into the engine control computer. This test is either part of a routine self-diagnostic operating mode or must be activated by the technician.

On some computer-controlled engines, certain components must be disconnected before attempting the power balance test. Because of the wide variations from manufacturer to manufacturer, always check the appropriate service manual.

Be careful not to run the engine with a shorted cylinder for more than 15 seconds. The unburned fuel in the exhaust can build up in the catalytic converter and create an unsafe situation. Also run the engine for at least 10 seconds between cylinder shortings.

Oil Pressure Gauge

Checking the engine's oil pressure will give you information about engine wear and the condition of the oil pump, pressure regulator, and the entire lubrication system. Lower than normal oil pressures can be caused by excessive engine bearing clearances. Oil pressure is checked at the sending unit passage with an externally mounted mechanical oil pressure gauge. Various fittings are usually supplied with the oil pressure gauge to fit different openings in the lubrication system.

To get accurate results from the test, make sure you follow the manufacturer's recommendations and compare your findings to specifications. Low oil pressure readings can be caused by internal component wear, pump-related problems, low oil level, contaminated oil, or low oil viscosity. An overfilled crankcase, high oil viscosity, or a faulty pressure regulator can cause high oil pressure readings.

Stethoscope

Some engine sounds can be easily heard without using a listening device, but others are impossible to hear unless they are amplified. A stethoscope is very helpful in locating engine noise by amplifying the sound waves. It can also distinguish between normal and abnormal noise. The procedure for using a stethoscope is simple. Use the metal prod to trace the sound until it reaches its maximum intensity. Once the precise location has been discovered, the sound can be better evaluated. A sounding stick, which is nothing more than a long, hollow tube, works on the same principle, though a stethoscope gives much clearer results.

The best results, however, are obtained with an electronic listening device. With this tool you can tune into the noise. Doing this allows you to eliminate all other noises that might distract or mislead you.

Cooling System Pressure Tester

A cooling system pressure tester contains a hand pump and a pressure gauge. A hose is connected from the hand pump to a special adapter that fits on the radiator filler neck. This tester is used to pressurize the cooling system and check for coolant leaks. Additional adapters are available to connect the tester to the radiator cap. With the tester connected to the radiator cap, the pressure relief action of the cap may be checked.

Coolant Hydrometer

A coolant hydrometer is used to check the amount of antifreeze in the coolant. This tester contains a pickup hose, coolant reservoir, and squeeze bulb. The pickup hose is placed in the radiator coolant. When the squeeze bulb is squeezed and released, coolant is drawn into the reservoir. As coolant enters the reservoir, a pivoted float moves upward with the coolant level. A pointer on the float indicates the freezing point of the coolant on a scale located on the reservoir housing.

Chassis Dynamometer

A chassis dynamometer, commonly called a dyno, is used to simulate a road test. A vehicle can be driven through a wide assortment of operating conditions, without leaving the shop. Because the vehicle is stationary, test equipment can be connected and monitored while the vehicle is driven under various loads. This is extremely valuable when diagnosing a problem. A chassis dyno can also be used for performance tuning.

The vehicle's drive wheels are positioned on large rollers. The electronically-controlled rollers offer rotational resistance to simulate the various loads a vehicle may face.

Some performance shops have an engine dynamometer that directly measures the output from an engine. A chassis dynamometer measures the engine's output after it has passed through the driveline.

Service Information

Perhaps the most important tools you will use are service manuals. There is no way a technician can remember all of the procedures and specifications needed to correctly repair all vehicles. Therefore, a good technician relies on service manuals and other information sources for this information. Good information plus knowledge allows a technician to fix a problem with the least bit of frustration and at the lowest cost to the customer.

To obtain the correct engine specifications and other information, you must first identify the engine you are working on. The best source for engine identification is the VIN. The engine code can be interpreted through information given in the service manual. The manual may also help you identify the engine through casting numbers and/or markings on the cylinder block or head.

The primary source of repair and specification information for any car, van, or truck is the manufacturer. The manufacturer publishes service manuals each year, for every vehicle built. Because of the enormous amount of information, some manufacturers publish more than one manual per year per car model. They are typically divided into sections based on the major systems of the vehicle. In the case of engines, there is a section for each engine that may be found in the vehicle. Manufacturers' manuals detail all repairs, adjustments, specifications, diagnostic procedures, and special tools required.

Since many technical changes occur on specific vehicles each year, manufacturers' service information needs be constantly updated. Updates are published as service bulletins (often referred to as Technical Service Bulletins or TSBs) that show the changes in specifications and repair procedures during the model year. The car manufacturer provides these bulletins to dealers and repair facilities on a regular basis. Service information is updated on a regular basis on the Internet.

Service information is also published by independent companies rather than by the manufacturers. However, they pay for and get most of their information from the car makers. They contain component information, diagnostic steps, repair procedures, and specifications for several car makes in one book. Information is usually condensed and is more general in nature than the manufacturer's manuals. The condensed format allows for more coverage in less space and therefore is not always specific. They may also contain several years of models as well as several car makes in one book.

Many of the larger parts manufacturers have excellent guides on the various parts they manufacture or supply. They also provide updated service bulletins on their products. Other sources for up-to-date technical information are trade magazines and trade associations.

Service information is now commonly found electronically on digital video disks (DVDs), and on the Internet. A single DVD can hold a huge amount of text, eliminating the need for a library to contain all of the printed manuals. Using electronics to find information is also easier and quicker. The disks are normally updated quarterly, and not only contain the most recent service bulletins but also engineering and field service fixes. The current trend is to provide service information via the Internet. The Internet allows the information provider to instantly update the required files as changes or TSBs occur. All a technician needs to do is enter vehicle information and then move to the appropriate part or system. The appropriate information will then appear on the computer's screen. Online data can be updated instantly and requires no space for physical storage. These systems are easy to use and the information is quickly accessed and displayed. The computer's keyword, mouse, and/or light pen are used to make selections from the screen's menu. Once the information is retrieved, a technician can read it off the screen or print it out and take it to the service bay.

JOB SHEETS

REQUIRED SUPPLEMENTAL TASKS (RST)

JOB SHEET 1

Shop Safety Survey

Name ______________________ Station ______________ Date ______________

NATEF Correlation

This Job Sheet addresses the following **RST** tasks:

Shop and Personal Safety:

1. Identify general shop safety rules and procedures.
8. Identify the location and use of eyewash stations.
10. Comply with the required use of safety glasses, ear protection, gloves, and shoes during lab/shop activities
11. Identify and wear appropriate clothing for lab/shop activities
12. Secure hair and jewelry for lab/shop activities.

Objective

As a professional technician, safety should be one of your first concerns. This job sheet will increase your awareness of shop safety rules and safety equipment. As you survey your shop area and answer the following questions, you will learn how to evaluate the safeness of your workplace.

Materials

Copy of the shop rules from the instructor

PROCEDURE

Your instructor will review your progress throughout this worksheet and should sign off on the sheet when you complete it.

1. Have your instructor provide you with a copy of the shop rules and procedures.

   Have you read and understood the shop rules? ☐ Yes ☐ No

2. Before you begin to evaluate your work area, evaluate yourself. Are you dressed to work safely? ☐ Yes ☐ No

   If no, what is wrong? ______________________________

3. Are your safety glasses OSHA approved? ☐ Yes ☐ No

   Do they have side protection shields? ☐ Yes ☐ No

4. Look around the shop and note any area that poses a potential safety hazard.

   ____

   ____

   All true hazards should be brought to the attention of the instructor immediately.

5. What is the air line pressure in the shop? __________ psi

   What should it be? __________ psi

6. Where is the first-aid kit(s) kept in the work area?

   ____

   ____

7. Ask the instructor to show the location of, and demonstrate the use of, the eyewash station. Where is it and when should it be used?

   ____

   ____

8. What is the shop's procedure for dealing with an accident?

   ____

   ____

   ____

9. Explain how to secure hair and jewelry while working in the shop.

   ____

10. List the phone numbers that should be called in case of an emergency.

    ____

Problems Encountered

____

____

____

Instructor's Comments

____

____

____

JOB SHEET 2

Working in a Safe Shop Environment

Name ______________________ Station ______________ Date ______________

NATEF Correlation

This Job Sheet addresses the following **RST** tasks:

Shop and Personal Safety:

2. Utilize safe procedures for handling of tools and equipment.
3. Identify and use proper placement of floor jacks and jack stands.
4. Identify and use proper procedures for safe lift operation.
5. Utilize proper ventilation procedures for working within the lab/shop area.
6. Identify marked safety areas.
9. Identify the location of the posted evacuation routes.

Objective

This job sheet will help you work safely in the shop. Two of the basic tools a technician uses are lifts and the floor jack.

This job sheet also covers the technicians' environment.

Materials

Vehicle for hoist and jack stand demonstration

Service information

Describe the vehicle being worked on:

Year ________ Make ____________________ Model ____________________

VIN ____________________ Engine type and size ____________________

PROCEDURE

1. Are there safety areas marked around grinders and other machinery? ☐ Yes ☐ No
2. Are the shop emergency escape routes clearly marked? ☐ Yes ☐ No
3. Have your instructor demonstrate the exhaust gas ventilation system in the shop. Explain the importance of the ventilation system.

____

____

4. What types of Lifts are used in the shop?

____

____

5. Find the location of the correct lifting points for the vehicle supplied by the instructor. On the rear of this sheet, draw a simple figure showing where these lift points are.

6. Ask your instructor to demonstrate the proper use of the lift.

   Summarize the proper use of the lift.

____

____

____

____

7. Demonstrate the proper use of jack stands with the help of your instructor.

   Summarize the proper use of jack stands.

____

____

____

Problems Encountered

____

____

____

Instructor's Comments

____

____

____

JOB SHEET 3

Fire Extinguisher Care and Use

Name ______________________________ Station ________________ Date ________________

NATEF Correlation

This Job Sheet addresses the following **RST** task:

Shop and Personal Safety:

7. Identify the location and the types of fire extinguishers and other fire safety equipment; demonstrate knowledge of the procedures for using fire extinguishers and other fire safety equipment.

Objective

Upon completion of this job sheet, you will be able to demonstrate knowledge of the procedures for using fire extinguishers and other fire safety equipment, and identify the location of fire extinguishers in the shop.

NOTE: *Never fight a fire that is out of control or too large. Call the fire department immediately!*

1. Identify the location of the fire extinguishers in the shop.

   ____

   ____

2. Have the fire extinguishers been inspected recently? (Look for a dated tag.)

   ____

   ____

3. What types of fires are each of the shop's fire extinguishers rated to fight?

   ____

   ____

4. What types of fires should not be used with the shop's extinguishers?

   ____

   ____

5. One way to remember the operation of a fire extinguisher is to remember the term PASS. Describe the meaning of PASS on the following lines.

   a. P

   ____

   b. A

   ____

c. S

____

d. S

____

Problems Encountered

____

____

____

Instructor's Comments

____

____

____

JOB SHEET 4

Working Safely Around Air Bags

Name ______________________________ Station ________________ Date ________________

NATEF Correlation

This Job Sheet addresses the following **RST** task:

Shop and Personal Safety:

13. Demonstrate awareness of the safety aspects of supplemental restraint systems (SRS), electronic brake control systems, and hybrid vehicle high-voltage circuits.

Objective

Upon completion of this job sheet, you should be able to work safely around and with air bag systems.

Tools and Materials

A vehicle(s) with air bag

Safety glasses, goggles

Service information appropriate to vehicle(s) used

Describe the vehicle being worked on:

Year __________ Make ________________________ Model ________________________

VIN ________________________ Engine type and size ________________________

PROCEDURE

1. Locate the information about the air bag system in the service information. How are the critical parts of the system identified in the vehicle?

____

____

2. List the main components of the air bag system and describe their location.

____

____

3. There are some very important guidelines to follow when working with and around air bag systems. Look through the service information to find the answers to the questions and fill in the blanks with the correct words.

   a. Wait at least __________ minutes after disconnecting the battery before beginning any service. The reserve __________ module is capable of storing enough energy to deploy the air bag for up to __________ minutes after battery voltage is lost.

b. Never carry an air bag module by its __________ or __________, and, when carrying it, always face the trim and air bag __________ from your body. When placing a module on a bench, always face the trim and air bag __________.

c. Deployed air bags may have a powdery residue on them. __________ is produced by the deployment reaction and is converted to __________ when it comes in contact with the moisture in the atmosphere. Although it is unlikely that harmful chemicals will still be on the bag, it is wise to wear __________ and __________ when handling a deployed air bag. Immediately wash your hands after handling a deployed air bag.

d. A live air bag must be __________ before it is disposed. A deployed air bag should be disposed of in a manner consistent with the __________ and manufacturer's procedures.

e. Never use a battery- or AC-powered __________, __________, or any other type of test equipment in the system unless the manufacturer specifically says to. Never probe with a __________ for voltage.

4. Explain how an air bag sensor should be handled before it is installed on the vehicle.

Problems Encountered

____

____

____

Instructor's Response

____

____

____

JOB SHEET 5

High-Voltage Hazards in Today's Vehicles

Name ______________________ Station ______________ Date ______________

NATEF Correlation

This Job Sheet addresses the following **RST** task:

Shop and Personal Safety:

14. Demonstrate awareness of the safety aspects of high-voltage circuits (such as high intensity discharge (HID) lamps, ignition systems, injection systems, etc.).

Objective

Upon completion of this job sheet, you will be able to describe some of the necessary precautions to take in performing work around high-voltage hazards such as HID headlamps and ignition systems.

HID Headlamp Precautions

Describe the vehicle being worked on:

Year ________ Make ____________________ Model ____________________

VIN ____________________ Engine type and size ____________________

Materials needed

Service information for HID headlamps

Describe general operating condition:

1. List three precautions a technician should observe when working with HID headlamp systems.

____

____

____

2. How many volts are necessary to initiate and maintain the arc inside the bulb of this HID system?

____

____

3. A technician must never probe with a test lamp between the HID ballast and the bulb. Explain why?

____

High-Voltage Ignition System Precautions

4. High-voltage ignition systems can cause serious injury, especially for those who have heart problems. Name at least three precautions to take when working around high-voltage ignition systems.

____

____

____

____

Problems Encountered

____

____

____

Instructor's Comments

____

____

____

JOB SHEET 6

Hybrid High-Voltage and Brake System Pressure Hazards

Name ______________________ Station ______________ Date ______________

NATEF Correlation

This Job Sheet addresses the following **RST** task:

Shop and Personal Safety:

13. Demonstrate awareness of the safety aspects of supplemental restraint systems (SRS), electronic brake control systems, and hybrid vehicle high-voltage circuits.

Objective

Upon completion of this job sheet, you will be able to describe some of the necessary precautions to take while performing work around high-voltage systems such as that used in hybrid vehicles, as well as the high-pressure systems of some braking control systems.

Tools and Materials

Appropriate service information

AUTHOR'S NOTE: *According to the vehicles' manufacturers, a technician should not work on hybrid vehicles without having the specific training, which is beyond the scope of this job sheet. This job sheet assumes that the student will be accessing information only, not actually working on live high-voltage vehicles.*

Protective Gear

Goggles or safety glasses with side shields

High-voltage gloves with properly inspected liners

Orange traffic cones to warn others in the shop of a high-voltage hazard

Describe the vehicle being worked on:

Year ________ Make ____________________ Model ____________________

VIN ____________________ Engine type and size ____________________

Describe general operating condition:

____

PROCEDURE

High Pressure Braking Systems:

NOTE: *Many vehicles have a high pressure accumulator or a high pressure pump in their braking systems. Opening one of these systems can be hazardous due to the high pressures involved.*

1. Research the vehicle you have been assigned and describe the procedure that must be followed BEFORE opening the hydraulic braking system.

____

____

____

Hybrid Vehicles:

1. Special gloves with liners are to be inspected before each use when working on a hybrid vehicle.

   A. How should the integrity of the gloves be checked?

   ____

   ____

   B. How often should the gloves be tested (at a lab) for recertification?

   ____

   C. What precautions must be taken when storing the gloves?

   ____

   ____

   ____

   D. When must the gloves be worn?

   ____

   ____

   ____

2. What color are the high-voltage cables on hybrid vehicles?

____

3. What must be done BEFORE disconnecting the main voltage supply cable?

____

____

____

4. Describe the safety basis of the "one hand rule."

____

____

____

5. Explain the procedure to disable high voltage on the vehicle you selected.

____

____

____

6. Explain the procedure to test for high voltage to ensure that the main voltage is disconnected.

____

____

Problems Encountered

____

____

____

Instructor's Comments

____

____

____

JOB SHEET 7

Material Data Safety Sheet Usage

Name ______________________ Station ______________ Date ______________

NATEF Correlation

This Job Sheet addresses the following **RST** task:

Shop and Personal Safety:

12. Locate and demonstrate knowledge of material safety data sheets (MSDS).

Objective

On completion of this job sheet, the student will be able to locate the MSDS folder and describe the use of an MSDS sheet on the job site.

Materials

Selection of chemicals from the shop

MSDS sheets

1. Locate the MSDS folder in the shop. It should be in a prominent location. Did you have any problems finding the folder?

____

____

2. Pick a common chemical from your tool room such as brake cleaner. Locate the chemical in the MSDS folder. What chemical did you choose?

____

____

3. What is the flash point of the chemical? __________

4. Briefly describe why the flash point is important.

____

____

5. What is the first aid if the chemical is ingested?

____

____

6. Can this chemical be absorbed through the skin?

____

____

7. What are the signs of exposure to the chemical you selected?

___

___

8. What are the primary health hazards of the chemical?

___

___

9. What is the first aid procedure for exposure to this chemical?

___

___

10. What are the recommendations for protective clothing?

___

___

Problems Encountered

___

___

___

Instructor's Comments

___

___

___

JOB SHEET 8

Measuring Tools and Equipment Use

Name ______________________________ Station __________________ Date __________________

NATEF Correlation

This Job Sheet addresses the following **RST** tasks:

Tools and Equipment:

1. Identify tools and their usage in automotive applications.
2. Identify standard and metric designations.
3. Demonstrate safe handling and use of appropriate tools.
4. Demonstrate proper cleaning, storage, and maintenance of tools and equipment.
5. Demonstrate proper use of precision measuring tools (i.e., micrometer, dial-indicator, dial caliper).

Objective

Upon completion of this job sheet, you will be able to make measurements using micrometers, dial indicators, pressure gauges, and other measuring tools. You will also be able to demonstrate the safe handling and use of appropriate tools, and demonstrate their proper use.

Tools and Materials

Items to measure selected by the instructor

Precision measuring tools: micrometer (digital or manual), dial caliper, vacuum or pressure gauge (selected by the instructor)

PROCEDURE

Have your instructor demonstrate the measuring tools that are available to you in your shop. The tools should include both standard and metric tools.

1. Describe the measuring tools you will be using.

   ____

   ____

2. Describe the items that you will be measuring.

   ____

   ____

3. Describe any special handling or safety procedures that are necessary when using the tools.

   ____

   ____

4. Describe any special cleaning or storage that these tools might require.

____

____

5. Describe the metric unit of measurement of the tools, such as millimeters, centimeters, kilopascals, etc.

____

____

6. Describe the standard unit of measure of the tools, such as inches, pounds per square inch, etc.

____

____

7. Measure the components, list them, and record your measurements in the following table.

Item measured	*Measurement taken*	*Unit of measure*

8. Clean and store the tools. Describe the process next.

Problems Encountered

____

____

____

Instructor's Comments

____

____

____

JOB SHEET 9

Preparing the Vehicle for Service and Customer

Name ______________________ Station ______________ Date ______________

NATEF Correlation

This Job Sheet addresses the following **RST** tasks:

Preparing Vehicle for Service:

1. Identify information needed and the service requested on a repair order.
2. Identify purpose and demonstrate proper use of fender covers, mats.
3. Demonstrate use of the three C's (concern, cause, and correction).
4. Review vehicle service history.
5. Complete work order to include customer information, vehicle identifying information, customer concern, related service history, cause, and correction.

Preparing Vehicle for Customer:

1. Ensure vehicle is prepared to return to customer per school/company policy (floor mats, steering wheel cover, etc.).

Objective

Upon completion of this job sheet, you will be able to prepare a service work order based on customer input, vehicle information, and service history. The student will also be able to describe the appropriate steps to take to protect the vehicle and delivering the vehicle to the customer after the repair.

Tools and Materials

An assigned vehicle or the vehicle of your choice

Service work order or computer-based shop management package

Parts and labor guide

Work Order Source: Describe the system used to complete the work order. If a paper repair order is being used, describe the source.

____

____

PROCEDURE

1. Prepare the shop management software for entering a new work order or obtain a blank paper work order. Describe the type of repair order you are going to use.

 ____

 ____

2. Enter customer information, including name, address, and phone numbers onto the work order. Task Completed ☐

3. Locate and record the vehicle's VIN. Where did you find the VIN?

 ____

4. Enter the necessary vehicle information, including year, make, model, engine type and size, transmission type, license number, and odometer reading. Task Completed ☐

5. Does the VIN verify that the information about the vehicle is correct?

 ____

6. Normally, you would interview the customer to identify his or her concerns. However, to complete this job sheet, assume the only concern is that the customer wishes to have the front brake pads replaced. Also, assume no additional work is required to do this. Add this service to the work order. Task Completed ☐

7. Prepare the vehicle for entering the service department. Add floor mats, seat covers, and steering wheel covers to the vehicle. Task Completed ☐

8. The history of service to the vehicle can often help diagnose problems, as well as indicate possible premature part failure. Gathering this information from the customer can provide some of the data needed. For this job sheet, assume the vehicle has not had a similar problem and was not recently involved in a collision. Service history is further obtained by searching files for previous service. Often this search is done by customer name, VIN, and license number. Check the files for any related service work. Task Completed ☐

9. Search for technical service bulletins on this vehicle that may relate to the customer's concern. Did you find any? __________ If so, record the reference numbers here.

 ____

 ____

10. Based on the customer's concern, service history, TSBs, and your knowledge, what is the likely cause of this concern?

 ____

 ____

11. Add this information to the work order. Task Completed ☐

12. Prepare to make a repair cost estimate for the customer. Identify all parts that may need to be replaced to correct the concern. List these here.

____

____

13. Describe the task(s) that will be necessary to replace the part.

____

____

14. Using the parts and labor guide, locate the cost of the parts that will be replaced and enter the cost of each item onto the work order at the appropriate place for creating an estimate. If the valve or cam cover is leaking, what part will need to be replaced?

____

15. Now, locate the flat rate time for work required to correct the concern. List each task with its flat rate time.

____

____

16. Multiply the time for each task by the shop's hourly rate and add the cost of each item to the work order at the appropriate place for creating an estimate. Ask your instructor which shop labor rate to use and record it here.

____

17. Many shops have a standard amount they charge each customer for shop supplies and waste disposal. For this job sheet, use an amount of ten dollars for shop supplies. Task Completed ☐

18. Add the total costs and insert the sum as the subtotal of the estimate. Task Completed ☐

19. Taxes must be included in the estimate. What is the sales tax rate and does it apply to both parts and labor, or just one of these?

____

____

20. Enter the appropriate amount of taxes to the estimate, then add this to the subtotal. The end result is the estimate to give the customer. Task Completed ☐

21. By law, how accurate must your estimate be?

____

____

22. Generally speaking, the work order is complete and is ready for the customer's signature. However, some businesses require additional information; make sure you add that information to the work order. On the work order, there is a legal statement that defines what the customer is agreeing to. Briefly describe the contents of that statement.

____

23. Now that the vehicle is completed and it is ready to be returned to the customer, what are the appropriate steps to take to deliver the vehicle to the customer? What would you do to make the delivery special? What should not happen when the vehicle is delivered to the customer?

____

____

Problems Encountered

____

____

____

Instructor's Comments

____

____

____

JOB SHEETS

ENGINE PERFORMANCE JOB SHEET 10

Verifying the Condition of and Inspecting an Engine

Name ______________________ Station ______________ Date ______________

NATEF Correlation

This Job Sheet addresses the following **AST/MAST** tasks:

A.1. Identify and interpret engine performance concerns; determine necessary action.

A.3. Diagnose abnormal engine noises or vibration concerns; determine necessary action.

A.4. Diagnose the cause of excessive oil consumption, coolant consumption, unusual exhaust color, odor, and sound; determine necessary action.

Objective

Upon completion of this job sheet, you will be able to identify and interpret engine concerns; diagnose engine noises and vibrations; and diagnose the cause of excessive oil consumption, unusual engine exhaust color, odor, and sounds.

Tools and Materials

A road-worthy vehicle

Stethoscope

Protective Clothing

Goggles or safety glasses with side shields

Describe the vehicle being worked on:

Year ________ Make ____________________ Model ____________________

VIN ____________________ Engine type and size ____________________

PROCEDURE

Verify Engine Condition

1. Verifying the customer's concern is typically the first step you should take when diagnosing a problem. If the owner of the vehicle stated a concern, describe it. If there are no customer complaints, describe the general running condition to the best of your knowledge. The concern may be one of performance, smoke, leaks, or noises. In your answer, be sure to completely describe the condition and state when, where, and how the condition occurs.

____

____

____

____

2. Verify the complaint. Describe what you will do in an attempt to duplicate the concern. If necessary, road test the vehicle under the same conditions that are present when the problem normally occurs. Include in this description conditions that will explain when, where, and how the concern occurs.

____

____

____

____

3. Often, a customer may only notice poor performance during one condition, even though the problem may exist at others. Also, observing the performance of an engine during a variety of modes of operation may let you know what is working fine and what does not need to be tested further. Describe how the engine performed during the following conditions:

   A. Starting:

   ____

   B. Idling:

   ____

   C. Slow acceleration:

   ____

   D. Slow cruise:

   ____

   E. Slow deceleration:

   ____

   F. Heavy acceleration:

   ____

G. Highway cruise:

____

H. Fast deceleration:

____

I. Shut down:

____

4. Based on the results of the preceding checks, what engine systems do you think should be checked to find the cause of the customer's complaint?

____

____

5. If an engine uses excessive oil and there is no evidence of leaks, the oil may be burning in the combustion chambers. If excessive amounts of oil are burned in the combustion chambers, the exhaust contains blue smoke, and the spark plugs may be fouled with oil. Excessive oil burning in the combustion chambers may be caused by worn rings and cylinders or worn valve guides and valve seals. Remove the spark plugs and describe the condition of each.

#1 ____

#2 ____

#3 ____

#4 ____

#5 ____

#6 ____

#7 ____

#8 ____

6. A loss of coolant with no visible leaks can indicate an internal coolant leak. A whitish exhaust may be indicative of this sort of coolant leak if it is caused by a leaking head gasket or cracked engine components. Take a look at the exhaust while the engine is idling and describe it.

____

____

Engine Exhaust Diagnosis

1. Some engine problems may be diagnosed by the color, smell, or sound of the exhaust. If the engine is operating normally, the exhaust should be colorless. In severely cold weather, it is normal to see a swirl of white vapor coming from the tailpipe, especially when the engine and exhaust system are cold. This vapor is moisture in the exhaust, which is a normal by-product of the combustion process. Carefully observe the exhaust from

your test vehicle and describe what you see and/or smell during the following operating conditions:

A. Cold start-up:

____

B. Cold idle:

____

C. Warm start-up:

____

D. Warm idle:

____

E. Snap-throttle open:

____

F. Snap-throttle closed:

____

2. Based on the preceding observations, what are your conclusions? (Use the explanations that follow to guide your thoughts.)

____

____

If the exhaust is blue, excessive amounts of oil are entering the combustion chamber, and this oil is being burned with the fuel. When the blue smoke in the exhaust is more noticeable on deceleration, the oil is likely getting past the rings into the cylinder. If the blue smoke appears in the exhaust immediately after a hot engine is restarted, the oil is likely leaking down the valve guides.

- If black smoke appears in the exhaust, the air–fuel mixture is too rich. A restriction in the air intake, such as a plugged air filter, may be responsible for a rich air–fuel mixture.
- Gray smoke in the exhaust may be caused by coolant leaking into the combustion chambers. This may be most noticeable when the engine is first started after it has been shut off for over half an hour.
- A strong sulfur smell in the exhaust indicates a rich air–fuel mixture.

3. You may have noticed a change in sound during the preceding test. If you did, describe the sound change and operating mode in which the sound changed.

____

____

____

____

4. Use these guidelines to determine the possible cause of the sound. Then, state your best guess for the cause of the noise.
   - When the engine is idling, the exhaust from the tailpipe should have a smooth, even sound.
   - If, during idle, the exhaust emits a "puff" sound at regular intervals, a cylinder may be misfiring.
   - When this sound is present, check the engine's ignition and fuel systems, and the engine's compression.
   - If the vehicle makes excessive exhaust noise while the engine is accelerated, check the exhaust system for leaks.
   - A small exhaust leak may cause a whistling noise when the engine is accelerated.
   - If the exhaust system produces a rattling noise when the engine is accelerated, check the muffler and catalytic converter for loose internal components.
   - When the engine makes a wheezing noise at idle or while the engine is running at a higher rpm, check for a restricted exhaust system.

   ____

   ____

   ____

Engine Noise Diagnosis

1. Sounds from the engine itself can help you locate engine problems or help you identify a weakness in the engine before it becomes a big problem. Long before a serious engine failure occurs, the engine usually makes warning noises. Engine defects such as damaged pistons, worn rings, loose piston pins, worn crankshaft bearings, worn camshaft lobes, and loose and worn valve train components usually produce their own peculiar noises. Certain engine defects also cause a noise under specific engine operating conditions. Since it is sometimes difficult to determine the exact location of an engine noise, a stethoscope may be useful. The stethoscope probe is placed on, or near, the suspected component, and the ends of the stethoscope are installed in your ears. The stethoscope amplifies sound to assist in noise location. When the stethoscope probe is moved closer to the source of the noise, the sound is louder in your ears. If a stethoscope is not available, what can be safely used to amplify the sound and help locate the source of the noise?

   ____

CAUTION: *When placing a stethoscope probe in various locations on a running engine, be careful not to catch the probe or your hands in moving components such as cooling fan blades and belts.*

2. Since a lack of lubrication is a common cause of engine noise, always check the engine oil level and condition prior to noise diagnosis. Carefully observe the oil for contamination by coolant or gasoline. Check the oil level and condition on your test vehicle, then record your findings.

   ____

   ____

3. During the diagnosis of engine noises, always operate the engine under the same conditions as those that are present when the noise ordinarily occurs. Remember that aluminum engine components such as pistons expand more when heated than cast-iron alloy components do. Therefore, a noise caused by a piston defect may occur when the engine is cold but might disappear when the engine reaches normal operating temperature. If the customer has an engine noise concern, describe the noise and state when it occurs.

____

____

____

4. Duplicate the condition at which the noise typically occurs and describe all that you hear as you listen to the engine.

____

____

5. If you verified the customer's concern, use a stethoscope to find the spot where the noise is the loudest. Describe where this is.

____

____

6. What could be causing the noise to be loud at that spot?

____

____

7. To help you understand and use noise as a diagnostic tool, you will be given a description of an engine noise. Using your knowledge and any resources you have handy (such as your textbook), identify the conditions or problems that would cause each of the following noises:

   A. A hollow, rapping noise that is most noticeable on acceleration with the engine cold. The noise may disappear when the engine reaches normal operating temperature.

   ____

   B. A heavy thumping knock for a brief time when the engine is first started after it has been shut off for several hours. This noise may also be noticeable on hard acceleration.

   ____

   C. A sharp, metallic, rapping noise that occurs with the engine idling.

   ____

   D. A thumping noise at the back of the engine.

   ____

E. A rumbling or thumping noise at the front of the engine, possibly accompanied by engine vibrations. When the engine is accelerated under load, the noise is more noticeable.

____

F. A light, rapping noise at speeds above 35 mph (21 kph). The noise may vary from a light to a heavier rapping sound, depending on the severity of the condition. If the condition is very bad, the noise may be evident when the engine is idling.

____

G. A high-pitched clicking noise is noticeable in the upper cylinder area during acceleration.

____

H. A heavy clicking noise is heard with the engine running at 2,000 to 3,000 rpm. When the condition is severe, a continuous, heavy clicking noise is evident at idle speed.

____

I. A whirring and light rattling noise when the engine is accelerated and decelerated. Severe cases may cause these noises at idle speed.

____

J. A light clicking noise with the engine idling. This noise is slower than the piston or connecting rod noise and is less noticeable when the engine is accelerated.

____

K. A high-pitched clicking noise that intensifies when the engine is accelerated.

____

L. A noise that is similar to marbles rattling inside a metal can. This noise usually occurs when the engine is accelerated.

____

Problems Encountered

____

____

____

Instructor's Comments

____

____

____

ENGINE PERFORMANCE JOB SHEET 11

Gathering Vehicle Information

Name ______________________ Station ______________ Date ______________

NATEF Correlation

This Job Sheet addresses the following **MLR** task:

A.1. Research applicable vehicle and service information, vehicle service history, service precautions, and technical service bulletins.

This Job Sheet addresses the following **AST/MAST** task:

A.2. Research applicable vehicle and service information, vehicle service history, service precautions, and technical service bulletins.

Objective

Upon completion of this job sheet, you will be able to gather service information about a vehicle and its engine and related systems.

Tools and Materials

Appropriate service information

Computer

Protective Clothing

Goggles or safety glasses with side shields

Describe the vehicle being worked on:

Year ________ Make ____________________ Model ____________________

VIN ____________________ Engine type and size ____________________

PROCEDURE

1. Using the service information or other source, describe what each letter and number in the VIN for this vehicle represents.

____

____

____

____

2. While looking in the engine compartment, did you find a label regarding the specifications for the engine? Describe where you found it.

____

____

____

____

3. Summarize the information contained on this label.

____

____

____

____

4. Using the service information, locate the information about the vehicle's engine. List the major components of the system and describe the primary characteristics of the engine (number of valves, shape/configuration, firing order, etc.).

____

____

____

____

____

5. Describe the engine's support systems (fuel, air, ignition, exhaust, and emissions) and their control system(s).

____

____

____

____

____

6. Using the service information, locate and record all service precautions regarding working on the engine and its systems as noted by the manufacturer.

____

____

____

____

7. Using the information that is available, locate and record the vehicle's service history.

____

____

____

____

8. Using the information sources that are available, summarize all Technical Service Bulletins for this vehicle that relate to the engine and its systems.

____

____

____

____

Problems Encountered

____

____

____

Instructor's Comments

____

____

____

ENGINE PERFORMANCE JOB SHEET 12

Using a Vacuum Gauge

Name ______________________________ Station __________________ Date __________________

NATEF Correlation

This Job Sheet addresses the following **MLR** task:

A.2. Perform engine absolute (vacuum/boost) manifold pressure tests; determine necessary action.

This Job Sheet addresses the following **AST/MAST** task:

A.5. Perform engine absolute (vacuum/boost) manifold pressure tests; determine necessary action.

Objective

Upon completion of this job sheet, you will be able to properly perform engine absolute (vacuum/boost) manifold pressure tests.

Tools and Materials

Vacuum gauge

Various lengths of vacuum hose

Tee hose fittings

Protective Clothing

Safety glasses

Describe the vehicle being worked on:

Year __________ Make _________________________ Model _________________________

VIN _________________________ Engine type and size _________________________

Describe the general operating condition.

____

____

Give a definition of vacuum.

____

____

PROCEDURE

1. Carefully look over the engine's intake manifold to identify vacuum hoses. Select one that is small and easily accessible. DO NOT disconnect it until you have the approval of your instructor. Describe its location.

   ____

2. With the engine off, disconnect the selected hose. Connect the Tee fitting to the end of the hose and connect another hose from the Tee fitting to the place where the vacuum hose was originally connected. Task Completed ☐

3. Connect the vacuum gauge to the remaining connection at the Tee fitting. Task Completed ☐

4. Start the engine and observe the vacuum gauge. Task Completed ☐

5. Describe the gauge reading and the action of the needle.

   ____

   ____

6. Quickly open the throttle and allow it to quickly close. Task Completed ☐

7. Describe the gauge reading and the action of the needle.

   ____

   ____

8. Turn off the engine and disconnect the vacuum gauge and hoses. Then, reconnect the engine's hose to its appropriate connector. Summarize what you observed.

   ____

   ____

   ____

Problems Encountered

____

____

____

Instructor's Comments

____

____

____

ENGINE PERFORMANCE JOB SHEET 13

Conduct a Cylinder Power Balance Test

Name ______________________ Station ______________ Date ______________

NATEF Correlation

This Job Sheet addresses the following **MLR** task:

A.3. Perform cylinder power balance test; determine necessary action.

This Job Sheet addresses the following **AST/MAST** task:

A.6. Perform cylinder power balance test; determine necessary action.

Objective

Upon completion of this job sheet, you will be able to conduct a cylinder power balance test and accurately interpret the results.

Tools and Materials

Scan tool or tune-up scope

Protective Clothing

Goggles or safety glasses with side shields

Describe the vehicle being worked on:

Year ________ Make ____________________ Model ____________________

VIN ____________________ Engine type and size ____________________

PROCEDURE

1. What is the firing order for this engine and how are the cylinders numbered?

____

____

____

2. Describe the general running condition of the engine.

____

____

____

3. Connect the scan tool or scope to the engine according to the instructions given with the equipment. Task Completed ☐

4. Describe the basic procedure for conducting a cylinder balance test using the equipment.

____

____

5. Conduct a cylinder power balance test on the engine, using proper testing procedures (make sure not to short a cylinder for more than a few seconds (to prevent damage to the catalytic converter), then record the results next:

Cylinder #	1	2	3	4	5	6	7	8
rpm loss	___	___	___	___	___	___	___	___

6. Describe what is indicated by the results of this test.

____

____

____

7. Briefly explain what is actually being measured by a cylinder power balance test.

____

____

____

8. Based on the test results, describe the condition of the engine. (Does it agree with your original description of the engine?)

____

____

____

9. How and why would the readings be different if the camshaft intake lobes for the number two cylinder were severely worn?

____

____

____

Problems Encountered

____

____

____

Instructor's Comments

____

____

____

ENGINE PERFORMANCE JOB SHEET 14

Perform Cylinder Cranking and Running Compression Tests

Name ______________________ Station ______________ Date ______________

NATEF Correlation

This Job Sheet addresses the following **MLR** task:

A.4. Perform cylinder cranking and running compression tests; determine necessary action.

This Job Sheet addresses the following **AST/MAST** task:

A.7. Perform cylinder cranking and running compression tests; determine necessary action.

Objective

Upon completion of this job sheet, you will be able to perform engine cranking and running compression tests, evaluate the condition of an engine, and determine the services that are required.

Tools and Materials

Spark plug socket, extensions, and ratchet
Compression tester
Fender covers
Jumper wire
Air blower
Small oil can with flexible nozzle
Service information

Protective Clothing

Safety goggles or glasses with side shields

Describe the vehicle being worked on:

Year ________ Make ____________________ Model ____________________

VIN ____________________ Engine type and size ____________________

Describe the general operating condition:

____

____

PROCEDURE

Cranking Compression Test

1. Disable the ignition and fuel injection system. Task Completed ☐
2. Remove the spark plug cables from all spark plugs. Use an air blower to clean around the spark plugs. Remove all of the spark plugs and place them on a clean surface in the order in which they are removed from the engine. Task Completed ☐

3. Remove the air cleaner from the car. Block the throttle in its wide-open position. Task Completed ☐

4. Install the compression testing adapter (if necessary) into the number 1 spark plug hole, and connect the compression tester to it. Task Completed ☐

5. Crank the engine through four compression strokes and record the reading from the compression gauge on the chart at the end of this job sheet. Task Completed ☐

6. Remove the compression adapter from the spark plug hole. Squirt a small amount of oil into the cylinder through the spark plug hole. Task Completed ☐

7. Repeat steps 4 and 5 on the same cylinder. Task Completed ☐

8. Complete steps 4, 5, 6, and 7 for each of the remaining cylinders in the engine. Task Completed ☐

9. Clean each spark plug and replace each spark plug in the cylinder from where it was removed. Reconnect the spark plug cables. Remove whatever tool is blocking the throttle plate open, and reinstall the air cleaner. Task Completed ☐

10. Start the engine and check its operation to determine if everything was connected back properly. Task Completed ☐

Name ______________________ Station ______________ Date ______________

REPORT SHEET ON COMPRESSION TEST

Compression for Each Cylinder *Cylinder No.*	*Dry*	*Wet*
1.		
2.		
3.		
4.		
5.		
6.		
7.		
8.		

Conclusions and Recommendations ______________________________________

____

____

Running Compression Test

1. A running compression test can be done on all cylinders or the one or ones that have been identified as having a problem. Have you identified the bad cylinders in this engine? If so, what test(s) did you use to identify it?

____

____

2. Briefly explain the advantage a running compression test has over a cranking compression test.

____

____

3. Start the engine to allow it to slightly warm up. Make sure it does not get too hot to safely work on it. Task Completed ☐

4. Remove one spark plug (test the bad cylinder, if you know where it is). Ground the spark plug wire with a jumper wire to prevent module damage. What cylinder are you checking?

____

____

5. Disconnect the injector to that cylinder on a port-type fuel injection system. Task Completed ☐

6. Choose the correct compression hose adapter for the compression tester. Task Completed ☐

7. Make sure the seal is on the end of the hose and tightly install the hose into the spark plug bore. Task Completed ☐

8. Attach the compression gauge to the end of the hose. Task Completed ☐

9. Start the engine and allow it to idle. Watch the reading on the compression gauge. What was the reading? If you failed to get an accurate reading, relieve the pressure in the gauge (or remove the check valve in the end of the hose to allow the reading to change with engine speed) and check the pressure again.

____

____

10. Using the throttle linkage, quickly open and close the throttle while observing the compression gauge. What was your reading?

____

____

11. Turn off the engine. Task Completed ☐

12. What are your conclusions from this test? Unless the service information tells you otherwise, compression pressure at idle should be about half of cranking compression pressure or 50–75 psi. The pressure should increase to about 80% of cranking pressure (80 to 120 psi) when the throttle is snapped open.

____

____

13. Repeat the steps on all cylinders you desire to test. Record the results of the tests.

____

____

14. Was there more than a 20-psi difference between cylinders?

____

____

15. What do you know about this engine after you have completed the test?

____

____

Problems Encountered

____

____

____

Instructor's Comments

____

____

____

ENGINE PERFORMANCE JOB SHEET 15

Perform a Cylinder Leakage Test

Name ______________________ Station ______________ Date ______________

NATEF Correlation

This Job Sheet addresses the following **MLR** task:

A.5. Perform cylinder leakage test; determine necessary action.

This Job Sheet addresses the following **AST/MAST** task:

A.8. Perform cylinder leakage test; determine necessary action.

Objective

Upon completion of this job sheet, you will be able to properly perform cylinder leakage tests.

Tools and Materials

Breakover bar
Chalk
Compressed air
Fender covers
Indicator light
Jumper lead
Leakage tester and whistle
Radiator coolant (if applicable)
Screwdrivers
Service information
Socket set
Spark plug socket and ratchet
TDC indicator
Test adapter hose

Protective Clothing

Safety goggles or glasses with side shields

Describe the vehicle being worked on:

Year ________ Make ____________________ Model ____________________

VIN ____________________ Engine type and size ____________________

Describe general condition:

____

____

PROCEDURE

CAUTION: *Very high voltages are present with high-energy ignition systems. Do NOT use this procedure on cars with distributorless ignition systems unless the ignition system is totally disabled.*

1. Check the coolant level and fill, if needed. Run the engine until it reaches normal operating temperature. Then, turn it off and disable the ignition or fuel injection system. Task Completed ☐

2. Disconnect the spark plug cables from the plugs. Use compressed air to clean all foreign matter out of the plug wells. Remove all spark plugs. Set the plugs on a workbench or other clean surface in the order in which they were removed. Remove all plug gaskets or tubes from the cylinder head. Task Completed ☐

3. Remove the air cleaner. Block the throttle plate in a wide-open position using a screwdriver or similar tool. Disconnect the PCV hose from the crankcase. Task Completed ☐

4. Install the test adapter hose in the number 1 cylinder spark plug hole. Connect the tester whistle to the adapter hose. Task Completed ☐

5. Using a socket and ratchet on the crankshaft pulley nut or bolt, slowly rotate the engine in its normal direction until the test whistle sounds, indicating the beginning of the compression stroke. Remove the tester whistle from the adapter hose. Task Completed ☐

6. **NOTE:** *If the engine is not exactly at TDC when the air pressure from the tester is applied, the engine might rotate. Make sure to stay clear of moving pulleys and belts. If the vehicle has a manual transmission, make sure it is out of gear, and on any vehicle make sure the drive wheels are blocked and the parking brake is applied.*

7. Set up the cylinder leakage tester according to the manufacturer's instructions. Connect the tester to the adapter hose. Does the gauge show more than 20 percent leakage? If so, look for air leaking from the throttle plate, tailpipe, or crankcase and for air bubbles in the radiator. Record your results on the Report Sheet on Cylinder Leakage Test found at the end of this job sheet. Task Completed ☐

8. Disconnect the tester from the adapter hose. Resume rotating the engine using the socket and ratchet on the crankshaft nut or bolt until the next appropriate cylinder mark lines up with the chalk mark on the engine. It may be necessary to use the whistle to locate the TDC. Task Completed ☐

9. Remove the adapter from the previously tested cylinder and install it in the plug hole of the next cylinder in the firing order. Task Completed ☐

10. Repeat steps 6, 7, and 8 on each remaining cylinder. Task Completed ☐

Problems Encountered

____

____

____

Instructor's Comments

____

____

____

Name ______________________ Station ________________ Date ________________

REPORT SHEET ON CYLINDER LEAKAGE TEST		
No.	*Percentage*	*Leakage From*
1.		
2.		
3.		
4.		
5.		
6.		
7.		
8.		

Conclusions and Recommendations ____

____

____

ENGINE PERFORMANCE JOB SHEET 16

Diagnosing Engine Performance Problems

Name ______________________ Station ______________ Date ______________

NATEF Correlation

This Job Sheet addresses the following **AST/MAST** task:

A.9. Diagnose engine mechanical, electrical, electronic, fuel, and ignition concerns; determine necessary action.

Objective

Upon completion of this job sheet, you will be able to diagnose an engine's mechanical, electrical, electronic, fuel, and ignition concerns.

Tools and Materials

Service information for the following vehicle

An engine diagnostic scope or DSO with adapters

A diagnostic scan tool for the following vehicle

Protective Clothing

Goggles or safety glasses with side shields

Describe the vehicle being worked on:

Year ________ Make ____________________ Model ____________________

VIN ____________________ Engine type and size ____________________

Firing order ____________________

Describe general operating condition. Are there any obvious problems, such as a misfire, MIL light illuminated, etc.?

____

____

PROCEDURE

1. Connect the scan tool to the vehicle. Are there any codes current or history?

   ____

2. If there are codes, list them here.

   ____

   ____

   ____

3. If codes were present, how would you categorize the possible causes? Could they be mechanical, electrical, fuel-, or ignition-related? Explain your reasoning.

____

____

____

4. With the scan tool still connected, look at the short- and long-term fuel trim percentages and record your results here.

____

____

____

5. Remove a small vacuum line with the engine running. Watch the effect on the fuel trim numbers. Record your results here.

____

____

____

6. Reconnect the vacuum line and again watch and record your numbers here.

____

____

____

7. Turn off the engine and unplug an injector connector (if accessible) start the engine and record the fuel trim numbers. Explain what happened.

____

____

____

8. Did the MIL come on with the injector disconnected?

____

____

____

9. If so, which code set?

____

10. What could be a possible cause of an engine misfire on a single cylinder?

    engine mechanical ___________________________________

    electrical/electronic ________________________________

    ignition system ____________________________________

    fuel system _______________________________________

11. List the possible causes of a misfire on multiple cylinders?

    engine mechanical ____

    electrical/electronic ____

    ignition system ____

    fuel system ____

12. If your shop has the capability to scope the ignition system, connect the ignition scope to the engine. Start the engine and observe the height of the firing line at each cylinder. Are they within 3 kV of each other?

    ____

13. If they are not within 3 kV of each other, which cylinders are most different from the rest? ____

14. Are the heights of the firing lines all below 11 kV? ________________

15. Are the heights of the firing lines above 8 kV? ________________

16. Based on these findings, what do you conclude?

    ____

    ____

    ____

    ____

    ____

    ____

Problems Encountered

____

____

____

Instructor's Comments

____

____

____

ENGINE PERFORMANCE JOB SHEET 17

Verifying Engine Temperature

Name ______________________ Station ______________ Date ______________

NATEF Correlation

This Job Sheet addresses the following **MLR** task:

A.6. Verify engine operating temperature; determine necessary action.

This Job Sheet addresses the following **AST/MAST** task:

A.10. Verify engine operating temperature; determine necessary action.

Objective

Upon completion of this job sheet, you will be able to verify the operating temperature of an engine and diagnose the cause of abnormal temperatures.

Tools and Materials

Scan tool

Service information

Pyrometer

Protective Clothing

Goggles or safety glasses with side shields

Describe the vehicle being worked on:

Year ________ Make ____________________ Model ____________________

VIN ____________________ Engine type and size ____________________

PROCEDURE

1. Engines are designed to run within a narrow range of temperatures. When they operate within that range, they are running efficiently. Typically, abnormal temperatures are only noticed by the owner when there is insufficient heat inside the vehicle during cold weather or when steam rolls out from under the hood. Why would an engine that is running at lower than normal temperatures be less efficient?

   ____

   ____

   ____

2. Name the part of the cooling system that has the responsibility for maintaining and regulating the temperature of the coolant.

____

____

3. Most vehicles built after 2002 have a monitor for the engine thermostat. What are the conditions for causing the monitor to set a trouble code? List your findings here.

____

____

4. Higher than normal operating temperatures will also affect the engine's efficiency. Name five problems that may cause the engine to run hot.

____

____

____

____

____

5. There are times when the engine seems to be running colder or hotter than normal but the temperature gauge reads normally. In these cases, the engine's coolant temperature should be measured and compared to the reading on the vehicle's temperature gauge. The temperature of an engine should also be verified when the activity of the powertrain computer seems to indicate that it is responding to a temperature that is different than what the temperature gauge and/or engine coolant temperature sensor indicates. Engine temperature is commonly measured with a pyrometer. Before measuring, look up the temperature specification for the engine's thermostat. The specification is:

____

The source of the specification was:

____

6. Operate the engine for 15 minutes at idle speed. Then, read the operating temperature reading on the scan tool. The reading is:

____

7. Read the temperature indicated on the thermometer. The reading measured on the pyrometer is:

____

8. Compare the reading with the specification and state the difference. Explain why it may be different.

____

____

____

Problems Encountered

Instructor's Comments

ENGINE PERFORMANCE JOB SHEET 18

Verifying Camshaft Timing

Name ______________________ Station ______________ Date ______________

NATEF Correlation

This Job Sheet addresses the following **AST/MAST** task:

A.11. Verify correct camshaft timing.

Objective

Upon completion of this job sheet, you will be able to verify camshaft timing according to the manufacturer's specifications.

Tools and Materials

Spark plug socket
Compression gauge
Remote starter switch
Flashlight
Miscellaneous hand tools
Large breaker bar
New rocker arm or camshaft cover gasket

Protective Clothing

Goggles or safety glasses with side shields

Describe the vehicle being worked on:

Year ________ Make ____________________ Model ____________________

VIN ____________________ Engine type and size ____________________

PROCEDURE

1. If the timing belt or chain has slipped on the camshaft sprocket, the engine may fail to start because the valves are not properly timed in relation to the crankshaft. (Sometimes the engine turns over faster than normal due to a lack of compression.) When the timing belt or chain has only slipped a few cogs on the camshaft sprocket, the engine has a lack of power, and fuel consumption is excessive. Task Completed ☐

2. What results of another engine performance test could suggest you need to physically verify camshaft timing.

____

____

____

____

3. To check the valve timing, begin by removing the spark plug from the number 1 cylinder.

____

____

____

____

4. Disable the ignition and fuel injection system. (See the service information.) Block the drive wheels and apply the emergency brake; manual transmissions should be in neutral. Task Completed ☐

5. Connect a remote control switch to the starter solenoid terminal and the battery terminal on the solenoid. Task Completed ☐

6. Place your thumb on top of the spark plug hole at cylinder #1. If this hole is not accessible, place a compression gauge in the opening. Task Completed ☐

7. Crank the engine until compression is felt at the spark plug hole. Then, slowly crank the engine until the timing mark lines up with the zero-degree position on the timing indicator. (If the engine does not use a timing indicator, you may have to remove the timing belt cover to verify camshaft-to-crankshaft timing.) The number 1 piston is now at TDC on the compression stroke. On many engines, the timing mark is on the crankshaft pulley, and the timing indicator is mounted above the pulley. Task Completed ☐

8. Slowly crank the engine for one revolution until the timing mark lines up with the zero-degree position on the timing indicator. The number 1 piston is now at TDC on the exhaust stroke. Task Completed ☐

9. Remove the rocker arm or camshaft cover and install a breaker bar and socket on the crankshaft pulley nut. Observe the valve action while rotating the crankshaft about 30 degrees before and after TDC on the exhaust stroke. In this crankshaft position, the exhaust valve should close a few degrees after TDC on the exhaust stroke, and the intake valve should open a few degrees before TDC on the exhaust stroke. Is this what you observed? Task Completed ☐

____

10. If the valves did not open properly in relation to the crankshaft position, the valve timing is not correct. What should you do to correct it?

____

____

____

____

11. If the timing was correct, reinstall the rocker arm or camshaft cover with a new gasket. Tighten the attaching bolts to the proper specification. The recommended torque is:

____

12. Reinstall the spark plug and tighten it to the proper specification. The recommended torque is:

____

Problems Encountered

____

____

____

Instructor's Comments

____

____

____

ENGINE PERFORMANCE JOB SHEET 19

Conduct a Diagnostic Check on an Engine Equipped with OBD II

Name ______________________ Station ______________ Date ______________

NATEF Correlation

This Job Sheet addresses the following **MLR/AST/MAST** task:

B.1. Retrieve and record diagnostic trouble codes, OBD monitor status, and freeze frame data; clear codes when applicable.

Objective

Upon completion of this job sheet, you will be able to retrieve and record stored OBD II diagnostic trouble codes, OBD monitor status, and freeze frame data, and clear codes when applicable.

Tools and Materials

A vehicle equipped with OBD II

Scan tool

Service information

Protective Clothing

Goggles or safety glasses with side shields

Describe the vehicle being worked on:

Year ________ Make ____________________ Model ____________________

VIN ____________________ Engine type and size ____________________

Describe general operating condition:

____

PROCEDURE

1. Describe the scan tool being used:

   Model ____

2. Check all electrical connections, including the battery and computer ground, for clean and tight connections. What did you find?

   ____

   ____

   ____

3. Check all vacuum lines and hoses, as well as the tightness of all attaching and mounting bolts in the induction system. What did you find?

___

___

4. Check for damaged air ducts. What did you find?

___

5. Check the ignition circuit, especially the secondary cables for signs of deterioration, insulation cracks, corrosion, and looseness. What did you find?

___

6. Are there any unusual noises or odors? If there are, describe them and tell what may be causing them.

___

7. Inspect all related wiring and connections at the PCM. What did you find?

___

___

___

8. Gather all pertinent information about the vehicle and the customer's complaint. This should include detailed information about any symptoms from the customer, a review of the vehicle's service history, published TSBs, and the information in the service information. Summarize your findings.

___

___

___

___

9. Are there any obvious problems with the engine, ignition system, or air–fuel system? If so, describe them here.

___

___

___

10. What are your conclusions from the preceding information?

___

___

___

11. Check the operation of the MIL by turning the ignition ON. Describe what happened and what it means.

____

____

12. Connect the scan tool to the DLC. Enter the vehicle identification information into the scan tool, and then retrieve the DTCs with the scan tool. List all codes retrieved by the scan tool.

____

____

____

____

13. What are your conclusions from these tests and checks?

____

____

____

14. Work through the menu on the scan tool until its screen reads "DATA LIST." Then, monitor the activity of the system and record any unusual activity. Describe your findings here.

____

____

____

15. Review any freeze frame data that may be related to the DTC or abnormal serial data. Are you able to identify the conditions present when the problem occurs?

____

____

16. What are your service recommendations?

____

____

____

17. After all codes have been recorded and repairs are made or planned, the DTCs should be cleared. This is done to verify that the service corrected the problem. Codes can be cleared with the scan tool. Make sure the scan tool is connected to the DLC. What did you find?

____

____

____

18. After the vehicle has been repaired, it is important to drive the vehicle until all the OBD II monitors have been run and completed. List the OBD II monitors that have not been run by the ECM at this time (before driving).

    ____

19. Describe the reason that it is important to make sure that all monitors have been run and completed before returning the vehicle to the customer.

    ____

    ____

    ____

Problems Encountered

____

____

____

Instructor's Comments

____

____

____

ENGINE PERFORMANCE JOB SHEET 20

Electronic Service Information

Name ______________________ Station ______________ Date ______________

NATEF Correlation

This Job Sheet addresses the following **AST/MAST** task:

B.2. Access and use service information to perform step-by-step (troubleshooting) diagnosis.

Objective

Upon completion of this job sheet, you will be able to access and use electronic service information (ESI).

Tools and Materials

Computer with ESI available

Protective Clothing

Goggles or safety glasses with side shields

Describe the vehicle being worked on:

Year ________ Make ____________________ Model ____________________

VIN ____________________ Engine type and size ____________________

PROCEDURE

1. Describe the computer and operating system you are using.

 ____

 ____

2. What is the source of the information you will be retrieving?

 ____

 ____

3. Enter into the software package the vehicle information requested. What did you need to enter?

 ____

 ____

4. What area of the vehicle do you want to retrieve information about, or what area was assigned to you?

 ____

 ____

5. Select that area on the computer. View the table of contents for that area and select the category you are researching. What is the category?

____

____

6. Select the section of that category you are researching. What is the section?

____

____

7. Select the type of information you want about this section. What do you want (TSBs, procedures, tips, etc.)?

____

____

8. Describe the procedure given in the service information for discovering the cause of a problem with the input from a throttle position sensor (normally code P0120).

____

____

Problems Encountered

____

____

____

Instructor's Comments

____

____

____

ENGINE PERFORMANCE JOB SHEET 21

Using a Scan Tool to Test Actuators

Name ______________________ Station ______________ Date ______________

NATEF Correlation

This Job Sheet addresses the following **AST/MAST** task:

B.3. Perform active tests of actuators using a scan tool; determine necessary action.

Objective

Upon completion of this job sheet, you will be able to check the operation of various actuators with a scan tool.

Tools and Materials

Service information

Scan tool

Vehicle with OBD II

Protective Clothing

Goggles or safety glasses with side shields

Describe the vehicle being worked on:

Year ________ Make ____________________ Model ____________________

VIN ____________________ Engine type and size ____________________

PROCEDURE

NOTE: *The actuators or computer outputs that can be tested directly with a scan tool varies with vehicle model and accessories the vehicle is equipped with. Refer to the service information whenever conducting these tests.*

1. What type of scan tool are you using?

____

____

2. Refer to the operating manual for the scan tool and list the actuators that can be controlled directly by the tool.

____

____

____

____

3. Which of these will you check? Also, describe the special procedures for checking that actuator.

____

____

____

4. Connect the scan tool to the DLC. Turn the ignition switch ON. Start the engine and allow it to warm up. Select the actuator from the ACTIVE TEST menu on the scan tool. Activate the actuator and listen or look for the specified changes. What did you observe?

____

____

____

5. Refer to the service information and list the actuators that can be monitored by observing the data stream. List them here.

____

____

____

6. Choose one of those actuators and list it here with the specified test conditions and results.

____

____

____

7. Select that data on the scan tool and observe its activity under the specified conditions. Describe your observations.

____

____

____

Problems Encountered

____

____

____

Instructor's Comments

____

____

____

ENGINE PERFORMANCE JOB SHEET 22

OBDII Monitors

Name ____________________________ Station ________________ Date ________________

NATEF Correlation

This Job Sheet addresses the following **MLR** task:

B.2. Describe the importance of running all OBDII monitors for repair verification.

This Job Sheet addresses the following **AST/MAST** task:

B.4. Describe the importance of running all OBDII monitors for repair verification.

Objective

Upon completion of this job sheet, you will be able to explain the importance of running and how to run OBDII monitors to verify a repair.

Tools and Materials

Service information

Scan tool

Protective Clothing

Goggles or safety glasses with side shields

Describe the vehicle being worked on:

Year __________ Make ________________________ Model ________________________

VIN ________________________ Engine type and size ________________________

PROCEDURE

1. Describe in simple terms what an OBDII monitor is.

____

____

2. How often is a continuous monitor run?

____

____

3. When can DTCs be set, as a result of a continuous monitor?

____

____

4. Using the service information, list the continuous monitors provided by the OBDII system in this vehicle.

____
____
____
____
____
____

5. Using the service information, list the non-continuous monitors provided by the OBDII system in this vehicle.

____
____
____
____
____
____

6. For each of the preceding items, describe when each will run.

____
____
____
____
____
____

7. Describe the enable criteria for the non-continuous monitors.

____
____
____
____

8. How do you know when the enable criteria are met for a particular monitor?

____
____

9. What DTCs will be set if the readiness monitor fails?

10. Connect a scan tool to the vehicle, and then check the status of the monitors and list them.

11. Why is it important to check all monitors after a system repair has been made?

Problems Encountered

Instructor's Comments

ENGINE PERFORMANCE JOB SHEET 23

Interpreting Codes from an Engine Control System

Name ______________________________ Station ________________ Date ________________

NATEF Correlation

This Job Sheet addresses the following **MAST** task:

B.5. Diagnose the causes of emissions or drivability concerns with stored or active diagnostic trouble codes; obtain, graph, and interpret scan tool data.

Objective

Upon completion of this job sheet, you will be able to diagnose the causes of emissions or drivability concerns resulting from the failure of the computerized engine controls with stored diagnostic trouble codes.

Tools and Materials

Scan tool

Digital multimeter

Service information

Protective Clothing

Goggles or safety glasses with side shields

Describe the vehicle being worked on:

Year __________ Make _________________________ Model _________________________

VIN _________________________ Engine type and size _________________________

PROCEDURE

1. Refer to the listing of codes given in the service information. Briefly describe how the codes are grouped by letter and number.

   ____

   ____

   ____

   ____

2. Describe what is indicated by the following codes:

   P0037: ____

   P0143: ____

   P0010: ____

   P0100: ____

   P2195: ____

3. Conduct all preliminary checks of the engine, electrical system, and vacuum lines. Did you find any problems? What were they?

4. Connect the scan tool to the DLC. Enter the vehicle information into the scan tool. Retrieve the DTCs with the scan tool. What are they?

5. Refer to the service information and locate the description of the DTCs retrieved from the computer. What do they signify?

6. For each of the codes, summarize what steps the manufacturer recommends for locating the exact cause of the problem.

7. Follow those steps and summarize what you found.

8. After correcting the problem, drive the vehicle through the necessary enable criteria to set the code. Recheck the DTCs to make sure you properly corrected the problem. What did you find?

____

____

____

____

Problems Encountered

____

____

____

Instructor's Comments

____

____

____

ENGINE PERFORMANCE JOB SHEET 24

No-Code Diagnostics

Name ______________________________ Station __________________ Date __________________

NATEF Correlation

This Job Sheet addresses the following **MAST** task:

B.6. Diagnose emissions or drivability concerns without stored diagnostic trouble codes; determine necessary action.

Objective

Upon completion of this job sheet, you will be able to diagnose emissions or drivability concerns when there are no stored diagnostic codes.

Tools and Materials

Service information

Digital multimeter

Lab scope

Wiring diagram for the vehicle

Protective Clothing

Goggles or safety glasses with side shields

Describe the vehicle being worked on:

Year __________ Make __________________________ Model __________________________

VIN __________________________ Engine type and size __________________________

PROCEDURE

1. Determining which part or area of a computerized engine control system is defective requires having a thorough knowledge of how the system works and following a logical troubleshooting process. Electronic engine control problems are usually caused by defective sensors and, to a lesser extent, output devices. The logical procedure in most cases is, therefore, to check the input sensors and wiring first, then the output devices and their wiring, and, finally, the computer. Most late-model computerized engine controls have self-diagnosis capabilities. A malfunction recognized by the computer is stored in the computer's memory as a trouble code. Stored codes can be retrieved and the indicated problem areas checked further. When there is a problem but no codes are stored, the cause of the problem is something not monitored by the computer or is

something working within the acceptable range as seen by the computer. List the components and systems that are monitored by the computer of the vehicle you are working on.

____

____

____

____

2. State the major problem you are diagnosing and describe the general condition of the vehicle.

____

____

3. Diagnostics should begin with a visual inspection. Check the condition of the air filter and related hardware around the filter. Summarize your findings.

____

____

____

____

4. Check the battery and its cables, and the vehicle's wiring harnesses, connectors, and charging system for loose, corroded, or damaged connections. Summarize your findings.

____

____

5. Make sure all vacuum hoses are connected and are not pinched or cut. Summarize your findings.

____

____

6. Check all sensors and actuators for signs of physical damage. Summarize your findings.

____

____

7. Based on the symptoms, describe the systems and components that you feel could be the cause of the problem if the cause was not discovered in the visual inspection.

____

____

8. All PCMs cannot operate properly unless they have good ground connections and the correct voltage at the required terminals. A wiring diagram for the vehicle being tested must be used for these tests. Back-probe the battery terminal at the PCM with the ignition switch off and connect a digital voltmeter from this terminal to ground. How many volts did you measure?

____

____

9. The voltage at this terminal should be 12 volts. If 12 volts are not available at this terminal, check the computer fuse and related circuit. Is the fuse good?

____

____

10. Turn on the ignition switch and connect the red voltmeter lead to the other battery terminals at the PCM with the black lead still grounded. The voltage measured at these terminals should also be 12 volts with the ignition switch on. How many volts did you measure?

____

11. If the specified voltage is not available, test the voltage supply wires to these terminals. These terminals may be connected through fuses, fuse links, or relays. If you needed to check these, what were the results?

____

____

12. Computer ground wires usually extend from the computer to a ground connection on the engine or battery. With the ignition switch on, connect a digital voltmeter from the battery ground to the computer ground. The voltage drop across the ground wires should be 30 millivolts or less. If the voltage reading is greater than that or more than that specified by the manufacturer, repair the ground wires or connection. Summarize the results of this check.

____

____

13. Check the negative and positive terminals and cables of the battery. Conduct a voltage drop test across each. Summarize the results of this check.

____

____

14. Conduct a voltage drop test for the grounds at five of the system's sensors and actuators. List the five components whose ground you checked and summarize the results of each check.

____

____

15. Check the ground of the same system components with a lab scope. Poor grounds can allow EMI or noise to be present on the reference voltage signal. This noise causes small changes in the voltage going to the sensor. Therefore, the output signal from the sensor will also have these voltage changes. The computer will try to respond to these small rapid changes, which can cause a drivability problem. Summarize the results of these checks.

____

____

16. The traces on the scope should have been flat. If noise is present, move the scope's negative probe to a known good ground. If the noise disappears, the sensor's ground circuit is bad or has resistance. If the noise is still present, the voltage feed circuit is bad or there is EMI in the circuit from another source, such as the AC generator. Summarize the results of these checks.

____

____

17. If a voltage trace has a large spike and the circuit is fitted with a resistor to limit noise, the resistor may be bad. Clamping diodes are used on devices like A/C compressor clutches to eliminate voltage spikes. If the diode is bad, a negative spike will result. Capacitors or chokes are used to control noise from a motor or generator. Test the suspected component and summarize the results of that check.

____

____

18. Most sensors and output devices can be checked with an ohmmeter. List the ones you suspect as being faulty and that you will test with an ohmmeter.

____

____

19. For those components, list the resistance specifications and your measurements. Then, state your conclusions.

____

____

____

____

20. Many sensors, output devices, and circuit wiring can be diagnosed by checking the voltage to and from them. List the ones you suspect as being faulty and that you will test with a voltmeter.

____

____

21. For those components, list the voltage specifications and your measurements. Then, state your conclusions.

___

___

22. The activity of sensors and actuators can be monitored with a lab scope. By watching their activity, you are doing more than testing them. Often problems elsewhere in the system will cause a device to behave abnormally. These situations are identified by the trace on a scope and by the technician's understanding of a scope and the device being monitored. List the ones you suspect as being faulty and that you will test with a lab scope.

___

___

23. Summarize the results of your lab scope checks, and then state your conclusions.

___

___

Problems Encountered

___

___

___

Instructor's Comments

___

___

___

ENGINE PERFORMANCE JOB SHEET 25

Checking Common Sensors

Name ______________________ Station ______________ Date ______________

NATEF Correlation

This Job Sheet addresses the following **MAST** task:

B.7. Inspect and test computerized engine control system sensors, powertrain/engine control module (PCM/ECM), actuators, and circuits using a graphing multimeter (GMM)/digital storage oscilloscope (DSO); perform necessary action.

Objective

Upon completion of this job sheet, you will be able to properly inspect and test computerized engine control system sensors, powertrain control modules (PCM), actuators, and circuits.

Tools and Materials

Digital multimeter

Protective Clothing

Goggles or safety glasses with side shields

Describe the vehicle being worked on:

Year ________ Make ____________________ Model ____________________

VIN ____________________ Engine type and size ____________________

Describe general operating condition:

____

PROCEDURE

ECT Sensors

1. Describe the location of the ECT sensor.

____

____

____

2. What color are the wires that are connected to the sensor?

____

____

3. Record the resistance specifications for a normal ECT sensor for this vehicle.

____

____

____

____

____

____

____

____

4. Disconnect the electrical connector to the sensor. Refer to the service information and identify what terminals to use to check the sensor's resistance. What are they?

____

____

5. Measure the resistance of the sensor. It was ________ ohms at approximately ___________°F.

6. What are your conclusions from the preceding test.

____

____

____

____

TP Sensors

1. Describe the type of lab scope you are using.

   Model ____________________

2. Connect the lab scope across the TP sensor. Describe how you connected the scope to the sensor.

____

____

3. With the ignition on, move the throttle from closed to fully open and then allow it to close slowly. Observe the trace on the scope while moving the throttle. Describe what the trace looked like.

____

____

____

4. Based on the waveform of the TP sensor, what can you determine about the sensor?

____

____

5. With a voltmeter, measure the reference voltage to the TP sensor. The reading should be __________ volts. It was _________ volts.

6. What is the output voltage from the sensor when the throttle is closed?

   ______ volts

7. What is the output voltage of the sensor when the throttle is opened?

   ______ volts

8. Move the throttle from closed to fully open and then allow it to close slowly. Describe the action of the voltmeter.

____

____

9. What are your conclusions from the preceding tests?

____

____

____

____

O_2 Sensors

1. Describe the type of digital multimeter and lab scope you will be using.

   Models ____

2. Connect the voltmeter between the O_2 sensor wire and ground. Backprobe the connector near the O_2 sensor to connect the voltmeter to the sensor signal wire. With the engine idling, record and describe the voltmeter readings.

____

____

3. What does this test tell you about the sensor?

____

____

4. Remove the sensor from the exhaust manifold. Connect the voltmeter between the sensor wire and the case of the sensor. Using a propane torch, heat the sensor element. Observe and record the voltmeter reading.

____

____

5. What are your conclusions from this test?

____

____

6. Backprobe the sensor signal wire at the computer and connect a digital voltmeter from the signal wire to ground with the engine idling.

   Record the voltmeter readings: ______________________

7. Connect the voltmeter from the sensor case to the sensor ground wire on the computer.

   Record the voltmeter readings: ______________________

8. What are your conclusions from these two tests?

____

____

____

9. Connect the scan tool to the DLC. Observe and record what happens to the voltage reading from the sensor.

____

____

____

10. How many cross counts were there? ______________________

11. What is your explanation of the preceding test?

____

____

____

12. Connect a lab scope to the sensor and observe the trace. Observe and record what happens to the voltage reading from the sensor.

13. How many cross counts were there? ______________________

14. What is your explanation of the preceding test?

____

____

____

MAP Sensors

1. Describe the type of digital multimeter and lab scope you will be using.

   Model ______________________________________

   If the MAP sensor produces an analog voltage signal, follow this procedure.

2. With the ignition switch on, backprobe the 5-volt reference wire. Connect a voltmeter from the reference wire to ground. The reading is ______ volts.

3. If the reference wire is not supplying the specified voltage, what should be checked next?

____

4. With the ignition switch on, connect the voltmeter from the sensor ground wire to the battery ground. What is the measured voltage drop? ______ volts

5. What does this indicate?

____

____

6. Backprobe the MAP sensor signal wire and connect a voltmeter from this wire to ground with the ignition switch on. What is the measured voltage? ______ volts

7. What does this indicate?

____

____

8. How do you determine the barometric pressure based on these voltage readings?

____

____

____

____

9. Turn the ignition switch on and connect a voltmeter to the MAP sensor signal wire. Task Completed ☐

10. Connect a vacuum hand pump to the MAP sensor vacuum connection and apply 5 inches of vacuum to the sensor. What is the voltage reading? ______ volts

NOTE: *On some MAP sensors, the sensor voltage signal should change 0.7 to 1.0 volts for every 5 inches of vacuum change applied to the sensor. Always use the vehicle manufacturer's specifications. If the barometric pressure voltage signal was 4.5 volts with 5 inches of vacuum applied to the MAP sensor, the voltage should be 3.5 volts to 3.8 volts. When 10 inches of vacuum is applied to the sensor, the voltage signal should be 2.5 volts to 3.1 volts. Check the MAP sensor voltage at 5-inch intervals, from 0 to 25 inches. If the MAP sensor voltage is not within specifications at any vacuum, replace the sensor.*

11. Record the results of all vacuum checks.

____

____

12. What did these tests indicate?

____

____

13. Connect the scope to the MAP output and a good ground. Accelerate the engine and allow it to return to idle. Observe and describe the trace.

____

____

____

14. What did the trace show about the sensor?

____

____

____

Mass Air Flow Sensor

1. Using service information, determine which wires provide power, and which are the ground and signal wires. Backprobe and measure the voltage at each terminal with the key on. How do your results compare to the specifications from the service information? Record your results next.

   Power ____

   Ground ____

   Signal ____

2. MAF sensors send a varying frequency signal to the ECM, which corresponds to the mass of air flowing into the engine. Using a multimeter set to measure frequency, see what the reading is at the signal wire from the MAF sensor. Place the red lead on the signal wire and the black lead on a good ground. Turn the key on. The reading is ___________ Hz. Start the engine. Record the reading at idle: _______ Hz. Record the reading at 1,000 rpm: ___________ Hz

3. The MAF sensor can also be tested using a digital storage oscilloscope. Using a digital storage oscilloscope, take the same measurements that you made in step 2 and draw a simple representation of the waveforms here.

   Key on engine off:

   Idle ___________

   1,000 rpm _________

4. Do the waveforms look consistent and uniform?

____

Problems Encountered

____

____

____

Instructor's Comments

____

____

____

ENGINE PERFORMANCE JOB SHEET 26

Diagnosing Related Systems

Name ______________________ Station ______________ Date ______________

NATEF Correlation

This Job Sheet addresses the following **MAST** task:

B.8. Diagnose drivability and emissions problems resulting from malfunctions of interrelated systems (cruise control, security alarms, suspension controls, traction controls, A/C, automatic transmissions, non-OEM installed accessories, or similar systems); determine necessary action.

Objective

Upon completion of this job sheet, you will be able to diagnose drivability and emissions problems resulting from failures of interrelated systems.

Tools and Materials

Service information

Protective Clothing

Goggles or safety glasses with side shields

Describe the vehicle being worked on:

Year __________ Make ______________________ Model ______________________

VIN ______________________ Engine type and size ______________________

PROCEDURE

NOTE: *Whenever there is a customer complaint that relates to how the car operates, always check the systems that could cause the problem before diving into the engine and its systems. This job sheet is divided into the systems that could cause drivability problems. There is no attempt made to cover all of the possible problems that will affect the way a vehicle operates. Rather, what is covered are the basic systems and components that are common causes of drivability problems.*

Clutches

1. A bucking or jerking sensation can be felt if the clutch slips or its splines to the transmission shaft are worn.
2. If the clutch does not release, the engine may be hard to start or perhaps will not crank at all unless the transmission is in neutral.
3. To check for clutch drag, start the engine, depress the clutch pedal completely, and shift the transmission into first gear. Do not release the clutch. Then, shift the transmission into neutral and wait 5 seconds before

attempting to shift smoothly into reverse. It should take no more than 5 seconds for the clutch disc, input shaft, and transmission gears to come to a complete stop after disengagement. This period is normal and should not be mistaken for clutch drag. Clutch drag would be made evident by gear clashing when the transmission was shifted into reverse.

4. An unbalanced clutch assembly may cause the engine to run rough. This same problem can be caused by a loose flywheel or pressure plate.

5. Slipping can be caused by a contaminated clutch disc or incorrect clutch pedal freeplay.

6. To verify that there is slippage, set the parking brake and disengage the clutch. Shift the transmission into third gear, and increase the engine speed to about 2000 rpm. Slowly release the clutch pedal until the clutch engages. The engine should stall immediately. If it does not stall within a few seconds, the clutch is slipping.

7. If any of these seem to be a likely cause of the engine performance problem, diagnose this system and summarize the results here.

____

____

____

____

Torque Converters

1. A bad torque converter may provide less torque multiplication and lead to customers complaining of a lack of power.

2. An unbalanced or loose torque converter and/or flex plate will cause a vibration that will make the engine seem to run rough or have a misfire.

3. If the converter clutch doesn't lock during cruising speeds, the customer will notice a decrease in fuel economy. Check the lockup circuit and the converter control circuit to determine why the clutch is not locking.

4. If the clutch locks too soon or remains locked throughout all engine speeds, there may be a lack of power and poor acceleration. If the clutch is locked while the engine is idling, the engine will stall. If this situation exists, the engine may also stall during deceleration.

5. If the converter locks too soon, or stays engaged as the vehicle slows down, there may be a surge or bucking.

6. If any of these seem to be a likely cause of the engine performance problem, diagnose this system and summarize the results here.

____

____

____

____

Automatic Transmissions

1. A transmission that shifts too early or too late will affect fuel consumption and overall performance. When the transmission shifts too early, the available power to the wheels decreases, causing poor acceleration. When the gears change later than normal, the customer may complain about noise or may notice that fuel consumption has increased.
2. If the transmission slips, there will be a lack of power, poor acceleration, and poor fuel economy. Sometimes when there is slippage, the engine will seem to lurch or buck once the gear is finally fully engaged. There is a power loss and a sudden engagement of the gear; this causes a surge of power.
3. The engine can seem to run rough if the transmission (automatic or manual) mounts are loose or faulty. This is also true of bad engine mounts. The mounts dampen the normal vibration of an engine. If the mounts are bad, little or no dampening takes place.
4. If any of these seem to be a likely cause of the engine performance problem, diagnose this system and summarize the results here.

____

____

____

____

Driveline

1. Driveline problems can set up a vibration that may appear as a rough-running engine, especially at cruising or steady vehicle speeds.
2. A failed U-joint or damaged drive shaft can exhibit a variety of symptoms. A clunk that is heard when the transmission is shifted into gear is the most obvious. You can also encounter unusual noise, roughness, or vibration.
3. To help differentiate a potential drive train problem from other common sources of noise or vibration, it is important to note the speed and driving conditions at which the problem occurs. As a general guide, a worn U-joint is most noticeable during acceleration or deceleration and is less speed sensitive than an unbalanced tire (commonly occurring in the 30 to 60 mph range) or a bad wheel bearing (more noticeable at higher speeds).
4. If any of these seem to be a likely cause of the engine performance problem, diagnose this system and summarize the results here.

____

____

____

____

Heating and Air Conditioning

1. Power loss when the air conditioning is running is normal. However, too much power loss suggests the need for further diagnosis of the compressor and/or compressor drive. There may also be a change in speed as the compressor is engaged and disengaged. Nothing is wrong; this is normal because of the immediate load put on the engine.

2. If this seems to be a likely cause of the engine performance problem, diagnose this system and summarize the results here.

____

____

____

____

Brake Systems

1. Failure of the brakes to release is often caused by a tight or misaligned connection between the power unit and the brake linkage. Broken pistons, diaphragms, bellows, or return springs can also cause this problem. To help pinpoint the problem, loosen the connection between the master cylinder and the brake booster. If the brakes release, the problem is caused by internal binding in the vacuum unit. If the brakes do not release, look for a crimped or restricted brake line or similar problem in the hydraulic system.

2. Proper adjustment of the master cylinder pushrod is necessary to ensure proper operation of the power brake system. A pushrod that is too long causes the master cylinder piston to close off the replenishing port, preventing hydraulic pressure from being released and resulting in brake drag. A pushrod that is too short causes excessive brake pedal travel and causes groaning noises to come from the booster when the brakes are applied.

3. Parking brakes can also be a cause of poor performance and bad fuel economy. If the parking brakes do not release, the rear brakes will drag. Even if the brakes are only slightly dragging, drivability will be affected.

4. If any of these seem to be a likely cause of the engine performance problem, diagnose this system and summarize the results here.

____

____

____

____

Wheels and Tires

1. Tires should be inflated to the recommended amount of pressure. The ideal amount of pressure is listed on the identification decal of the vehicle

and is given in the vehicle's owner's manual. Underinflated tires have more rolling resistance and therefore will cause fuel economy to decrease, as well as available power. Tires that are only a little low will still have a negative effect on fuel economy. Check the pressure in the tires and look at the tires for outside wear. If the tires are severely worn, tell the customer that they should be replaced and explain why.

2. If replacement tires have been installed on the vehicle and they are not the same size as the original equipment, fuel economy and engine performance will be affected. The change in tire diameter changes the overall gear ratio of the drivetrain. The overall drive gear ratio is determined by the gear ratios in the drivetrain and the circumference of the tire. The circumference of the tire determines how many times it will rotate during a mile of travel.

3. If any of these seem to be a likely cause of the engine performance problem, diagnose this system and summarize the results here.

____

____

____

____

Problems Encountered

____

____

____

Instructor's Comments

____

____

____

ENGINE PERFORMANCE JOB SHEET 27

No-Start Diagnosis (Cranks but does not start)

Name ______________________ Station ______________ Date ______________

NATEF Correlation

This Job Sheet addresses the following **AST/MAST** task:

C.1. Diagnose (troubleshoot) ignition system–related problems such as no-starting, hard starting, engine misfire, poor drivability, spark knock, power loss, poor mileage, and emissions concerns; determine necessary action.

Objective

Upon completion of this job sheet, you will be able to diagnose no-start problems.

Tools and Materials

Service information
Fuel pressure gauge
Test spark plug
Digital multimeter

Protective Clothing

Goggles or safety glasses with side shields

Describe the vehicle being worked on:

Year ________ Make ____________________ Model ____________________

VIN ____________________ Engine type and size ____________________

PROCEDURE

1. The cause of a no-start condition can be located in the electrical/electronic, air, fuel, or ignition system, or be determined by a lack of compression. Begin by observing the MIL. Does the MIL come on with the key on? (Note: Some MILs will turn off after a period of time.) What would you suspect if the MIL never came on in a no-start situation?

____

____

____

2. While you have the key on, check the fuel gauge. Does it indicate fuel? Yes or no? _______

3. Inspect the air filter and the ductwork. Check for blockages because an adequate amount of air must be supplied for the air–fuel mixture. Summarize your findings.

____

____

____

4. Check the fuel pressure at the fuel rail. A pressurized fuel supply of good quality must be delivered to the properly operating injectors. If there is no pressure, refer to the service information to determine if there is a switch or relay that will shut off the fuel pump if a collision or a severe jarring is detected by the switch. Summarize your findings.

____

____

____

5. Check for voltage signals to the injectors. The injectors must receive a trigger signal to inject the fuel. Summarize your findings.

____

____

6. What type of ignition does the vehicle have?

____

7. Connect a test spark plug to the coil secondary wire (or directly to the ignition coil boot if it is a coil-on plug design) and ground the spark plug case. Perform a spark intensity check on each coil with a test spark plug. Crank the engine and observe the spark plug when connected to each coil. A bright, snapping spark indicates that the secondary voltage output is good. If the spark is weak or if there is no spark, check the primary circuit, including the crankshaft and camshaft sensors. A weak or no spark condition can also be caused by an open ignition cable, bad coil, or bad ignition module. Summarize your findings.

____

____

____

WARNING: *Do not crank or run an EI-equipped engine with a spark plug wire completely removed from a spark plug. This action may cause leakage defects in the coils or spark plug wires. Always use a test spark plug.*

8. If the test plug did not fire, test the spark plug wires (if applicable) connected to that coil. If these wires are good, the coil is probably bad. When the test spark plug does not fire on any plug, connect a voltmeter from the input (battery) terminal on each coil pack to ground. With the ignition switch on, the voltmeter should read 12 volts. If the voltage is less than that, test the wire from the ignition switch to the coils and check the ignition switch. Summarize your findings.

____

____

____

9. If a crank or cam sensor fails, the engine will not start. Both of these sensor circuits can be checked with a voltmeter or DSO. If the sensors are receiving the correct amount of voltage and have good low-resistance ground circuits, their output should be a digital signal or display a pulsing voltmeter reading while the engine is cranking. If any of these conditions do not exist, the circuit needs to be repaired or the sensor needs to be replaced. Summarize your findings.

10. If the cause of the no-start has not been found, check a sample of the fuel, using a clean, safe container. Is the fuel contaminated? Does the fuel smell stale, or appear to be contaminated with water? Summarize your findings here.

11. If the cause of the no-start condition has not yet been found, check for proper compression. Summarize your findings here.

Problems Encountered

Instructor's Comments

ENGINE PERFORMANCE JOB SHEET 28

Testing Crankshaft and Camshaft Position Sensors

Name ______________________ Station ______________ Date ______________

NATEF Correlation

This Job Sheet addresses the following **AST/MAST** task:

C.2. Inspect and test crankshaft and camshaft position sensor(s); perform necessary action.

Objective

Upon completion of this job sheet, you will be able to properly inspect and test crankshaft and camshaft position sensor(s); and determine their condition and the recommended service.

Tools and Materials

Scan tool

DMM

Appropriate service information

Lab scope (DSO)

Protective Clothing

Goggles or safety glasses with side shields

Describe the vehicle being worked on:

Year ________ Make ______________________ Model ________________

VIN ______________________ Engine type and size ______________________

Type of ignition system ______________________

Describe general operating condition:

____

PROCEDURE

1. Does the engine have both a crankshaft position (CKP) sensor and a camshaft position (CMP) sensor? If not, what does it have?

____

____

2. Connect a scan tool to the vehicle and check for any CKP- and CMP-related DTCs. Record your findings.

____

____

3. What is indicated by the DTCs?

____

____

4. Check the status of the monitors to see if a CKP or CMP fault caused a monitor to be incomplete or to not run. What did you find?

____

____

5. Carefully check the wring, connectors, and physical condition of the CKP and surrounding areas. Describe them here.

____

____

6. Carefully check the wring, connectors, and physical condition of the CMP and surrounding areas. Describe them here.

____

____

7. Count the number of wires leading to the CKP to determine if it is a magnetic pulse generator or a Hall-effect sensor. What type of sensor is it?

____

____

8. Count the number of wires leading to the CMP to determine if it is a magnetic pulse generator or a Hall-effect sensor. What type of sensor is it?

____

____

Testing Magnetic Pulse Generator Sensors

1. Refer to the service information. What are the resistance specifications for the CKP sensor?

____

2. Refer to the service information. What are the resistance specifications for the CMP sensor?

____

3. Disconnect the wires to the CKP and measure the resistance of the sensor. What was your measurement?

____

4. Compare your measurement to specifications and state what is indicated.

____

____

5. Disconnect the wires to the CMP and measure the resistance of the sensor. What was your measurement?

____

____

6. Compare your measurement to specifications and state what is indicated.

____

____

7. What type of signal should a magnetic pulse sensor generate?

____

8. Connect the leads of the DSO to the CKP's leads. Make sure the scope is set to read low values. What did you set the scope to read?

____

____

9. Disable the ignition and/or injection system and crank the engine with the starter. Observe and describe the waveforms on the scope.

____

____

____

____

10. Does the waveform of the CKP show the same number of equally spaced pulses as the engine has cylinders, plus a double pulse signal?

____

____

11. What did the waveform tell you about the CKP?

____

____

____

____

12. Move the leads of the DSO to the CMP's leads. Make sure the scope is set to read low values. What did you set the scope to read?

____

____

13. Disable the ignition and/or injection system and crank the engine with the starter. Observe and describe the waveforms on the scope.

____

____

____

____

14. What did the waveform tell you about the CKP?

____

____

____

____

Testing Hall-Effect Sensors

1. What type of signal should a Hall-effect sensor generate?

____

____

2. Using the service information, identify the signal wire from the CKP sensor. What color is it?

____

____

3. Connect the leads of the DSO to the CKP's signal lead. Make sure the scope is set to read low values. What did you set the scope to read?

____

____

4. Disable the ignition and/or injection system and crank the engine with the starter. Observe and describe the waveforms on the scope.

____

____

____

____

5. Does the waveform of the CKP show the same number of equally spaced pulses as the engine has cylinders?

____

____

6. What did the waveform tell you about the CKP?

____

____

____

____

7. Some manufacturers recommend that camshaft timing be verified before checking the CMP. How would you do that?

____

____

8. Using the service information, identify the signal wire from the CMP sensor. What color is it?

____

9. Move the leads of the DSO to CMP's leads. Make sure the scope is set to read low values. What did you set the scope to read?

____

____

10. Disable the ignition and/or injection system and crank the engine with the starter. Observe and describe the waveforms on the scope.

____

____

____

____

11. What did the waveform tell you about the CMP? Is there any evidence of RFI or electrical noise?

____

____

Problems Encountered

____

____

____

Instructor's Comments

____

____

____

ENGINE PERFORMANCE JOB SHEET 29

Checking, Reprogramming, and Replacing PCM/ECMs

Name ______________________ Station ______________ Date ______________

NATEF Correlation

This Job Sheet addresses the following **AST/MAST** task:

C.3. Inspect, test, and/or replace ignition control module, powertrain/engine control module; reprogram as necessary.

Objective

Upon completion of this job sheet, you will be able to properly determine when a powertrain/engine control module (PCM or ECM) should be replaced or reprogrammed and be able to properly replace or reprogram one.

Tools and Materials

Scan tool

Personal computer or Internet connectors for the scan tool

Service information

Protective Clothing

Goggles or safety glasses with side shields

Describe the vehicle being worked on:

Year ________ Make ____________________ Model ______________

VIN __________________ Engine type and size __________________

Describe general operating condition:

____

____

PROCEDURE

NOTE: *PCM/ECMs should only be suspected as being the cause of a problem when all other possible systems and components are verified to be working properly.*

1. Typical computer-related DTCs only relate to a broad statement of an internal problem or communication problems. Are any PCM/ECM DTCs set? If so, which ones and what do they indicate?

____

____

2. PCM/ECM operation can be monitored by observing the output it sends to serve as a reference for one of the sensors. Using the service information, identify the correct pin for the TP sensor. What is it?

____

____

3. What should the reference voltage for the TP sensor be?

____

4. Carefully backprobe the connector at that pin and measure the voltage. What is it?

____

5. If the reference voltage is not correct, check the wiring before condemning the computer. What did you find?

____

____

6. State your conclusions about the PCM/ECM.

____

____

CAUTION: *Never use an ohmmeter to check a PCM/ECM, because it can damage the unit's circuitry.*

Replacing a PCM/ECM

1. Before replacing a PCM/ECM, use a scan tool to obtain the PROM identification number or the EEPROM calibration number of the original unit. What was the number?

____

____

2. Install a memory saver. Turn the ignition off and then disconnect the negative cable of the battery. Locate the PCM/ECM. Unbolt the unit from its brackets. Where was the unit located and what needed to be moved to gain access to it?

____

____

3. Identification information is normally stamped or printed on the computer. Use this data and the vehicle information to get the correct replacement. How is this computer identified?

____

____

4. Securely mount the replacement computer and make sure the electrical connector is fully seated. After installation, use a scan tool to make sure there are no communication errors. While doing this, what did you find?

____

____

NOTE: *While handling a PCM/ECM, take care to avoid damaging the computer through static electricity by following these guidelines:*

- *Refer to the service information to identify the coding or labeling used by the manufacturer to warn technicians that some components are sensitive to electrostatic discharge.*
- *Avoid touching the electrical terminals of the part, unless you are instructed to do so. It is good practice to keep your fingers off all electrical terminals, because the oil from your skin can cause corrosion.*
- *When you are using a voltmeter, always connect the negative meter lead first.*
- *Do not remove a part from its protective package until it is time to install the part.*
- *When handling any electronic part, always touch a known good ground before handling it. This should be often repeated and done more frequently after sliding across a seat, sitting down from a standing position, or walking a distance.*
- *Before removing the part from its package, ground yourself and the package to a known good ground on the vehicle. To do this, keep one hand on a chassis ground or wear an anti-static wrist strap with its wire connected to a good ground.*

EEPROM (Flash) Programming

1. Describe the reason why you are reprogramming the PCM/ECM.

____

____

2. Reprogramming is done by downloading the vehicle's information through a computer, a computerized engine analyzer, or a scan tool. The method can be direct programming with a computer, indirect programming with a scan tool and computer, or remote programming with the computer removed from the vehicle. What method will you be using? What equipment will you be using?

____

____

____

3. Before beginning the process, make sure the battery is fully charged. Recharge the battery if necessary but do not attempt to reprogram while the battery is connected to the charger (unless the charger is a special

type designed for use while programming). The voltage spikes from the charger can void the programming or damage the computer. What is the voltage of the battery and what are your recommendations?

____

____

____

4. To begin programming, the computer or scan needs to know the vehicle information. What did you need to enter?

____

____

5. Before reprogramming, check the date of the last program update on the scan tool. If the program is the latest version, there is no need to reprogram. If the computer has not been updated, reprogramming should be done. Do you really need to reprogram the computer? Why?

____

____

6. Move to the programming mode and follow the instructions given on the scan tool or computer. Select the updated program on the scan tool or computer. The programming will then begin. **NOTE:** During programming, do not make any connections until the programming is complete. Task Completed ☐

7. After programming is complete, turn the ignition switch to the position called for on the scan tool or computer. What did you need to do?

____

____

8. Verify proper system operation and make sure you programmed the correct program by observing the calibration number on the scan tool or computer. What did you find?

____

____

Problems Encountered

____

____

____

Instructor's Comments

____

____

____

ENGINE PERFORMANCE JOB SHEET 30

Inspecting and Testing an Ignition System

Name ______________________ Station ______________ Date ______________

NATEF Correlation

This Job Sheet addresses the following **MLR** task:

A.7. Remove and replace spark plugs; inspect secondary ignition components for wear and damage.

This Job Sheet addresses the following **AST/MAST** task:

C.4. Remove and replace spark plugs; inspect secondary ignition components for wear and damage.

Objective

Upon completion of this job sheet, you will be able to inspect and test the ignition system pick-up sensor or the triggering devices. You will also be able to inspect and test the ignition primary circuit wiring and its components.

Tools and Materials

Clean rag

Appropriate service information

DMM

Test light

DSO

Protective Clothing

Goggles or safety glasses with side shields

Describe the vehicle being worked on:

Year ________ Make ____________________ Model ____________________

VIN ____________________ Engine type and size ____________________

Ignition type ____________________

Describe general operating condition:

____

____

Describe the type of ignition system found on this vehicle:

____

____

PROCEDURE

Fill in the following sections, where applicable.

Basic Inspection (Coil-Over-Plug)

1. Inspect the wiring to the coil. What did you find?

 ____

 ____

2. Remove the coil from the valve cover. Task Completed ☐
3. Does the coil boot show signs of damage or being soaked with oil? ☐ Yes ☐ No
4. Is the boot brittle? ☐ Yes ☐ No
5. Are there any white or grayish powdery deposits? ☐ Yes ☐ No
6. Does the coil show signs of leakage in the coil towers? ☐ Yes ☐ No
7. Remove the coil boot. Do the coil towers show any signs of burning? ☐ Yes ☐ No
8. Does the vehicle have an externally mounted module? ☐ Yes ☐ No
9. Is the control module tightly mounted to a clean surface? (If applicable, some modules are part of the ECM.) ☐ Yes ☐ No
10. Are the electrical connections to the module corroded? ☐ Yes ☐ No
11. Are the electrical connections to the module loose or damaged? ☐ Yes ☐ No

Basic Inspection (With spark plug cables)

1. Are the spark plug cables pushed tightly into the coil and onto the spark plugs? ☐ Yes ☐ No
2. Do the secondary cables have cracks or signs of worn insulation? ☐ Yes ☐ No
3. Are the boots on the ends of the secondary wires cracked or brittle? ☐ Yes ☐ No
4. Are there any white or grayish powdery deposits on secondary cables? ☐ Yes ☐ No
5. Do the coils show signs of leakage in the coil towers? ☐ Yes ☐ No
6. Do the coil towers show any signs of burning? ☐ Yes ☐ No
7. Separate the coils and inspect the underside of the coil and the ignition module wires. Are the wires loose or damaged? ☐ Yes ☐ No
8. Are the connections of the primary ignition system wiring tight connections? ☐ Yes ☐ No
9. Is the control module tightly mounted to a clean surface? ☐ Yes ☐ No
10. Are the electrical connections to the module corroded? ☐ Yes ☐ No
11. Are the electrical connections to the module loose or damaged? ☐ Yes ☐ No

For all ignition systems

Record your summary of the visual inspection. Include what looked good, as well as what looked bad.

____

____

____

____

Check the following components:

1. Describe the general appearance of the coil.

   ____

   ____

2. Locate the resistance specifications for the ignition coil in the service information and record those here.

   NOTE: *Some COP ignition coils have built-in modules and resistance checks. Check with the service information you have for details.*

   The primary winding should have ________ ohms of resistance.

   The secondary winding should have ________ ohms of resistance.

3. Disconnect the leads connected to the coil being tested. Connect the ohmmeter from the negative (tach) side of the coil to its container or frame. Observe the reading on the meter. The reading was ________ ohms.

4. What does this indicate?

   ____

   ____

5. Connect the ohmmeter from the center tower of the coil to its container or frame. Observe the reading on the meter. The reading was ________ ohms. What does this indicate?

   ____

   ____

6. Connect the ohmmeter across the primary winding of the coil. The reading was ________ ohms. Compare this to specifications. What does this reading indicate?

   ____

   ____

7. Connect the ohmmeter across the secondary winding of the coil. The reading was ________ ohms. Compare this to specifications. What does this reading indicate?

   ____

   ____

8. Based on the preceding tests, what is your conclusion about the coil?

____

____

Ignition Switch

1. Turn the ignition key off and disconnect the wire connector at the ignition module. Does the system have an ignition module? If so, where is it located?

____

____

2. Look at the wiring diagram and identify the color of wires that represent the battery input, the output to the ignition system, and the output to the starting system. Describe them here.

____

____

3. Disconnect the S terminal of the starter solenoid that prevents the engine from cranking when the ignition is in the run position. Then, turn the ignition switch to the run position. With the test light, probe the power wire to the switch. Was there voltage? ☐ Yes ☐ No
4. Probe the power output of the switch. Was there voltage? ☐ Yes ☐ No
5. Check for voltage at the battery or positive terminal of an ignition coil. Was there voltage? ☐ Yes ☐ No
6. Turn the key to the start position and check for voltage at the start power wire connector at the module. Was there voltage? ☐ Yes ☐ No
7. Check for voltage at the bat terminal of the ignition coil. Was there voltage? ☐ Yes ☐ No
8. What can you conclude about the ignition switch after conducting these tests?

____

____

9. Turn the ignition switch to the off position. Identify the power feed to the ignition module(s). Describe the wire and its location.

____

____

10. Install a small straight pin or wire probe into the appropriate module's power wire. Then, connect the positive lead of the DMM to the straight pin and the negative lead to ground. Turn the ignition switch to the run position. Your voltage reading is ______.

11. Turn the ignition switch to the start position. Your voltage reading is ______.

12. From all of the tests, what are your conclusions about the ignition switch?

____

____

Secondary Ignition Wires

1. Does the system have separate secondary wires or is the system a COP-type where the ignition coil mounts directly over the spark plug?

____

____

2. If available, remove each secondary wire (one at a time) from the coil and spark plug. Measure the resistance of each wire from end to end. Record your readings here.

____

____

3. What is the specified resistance? __________ ohms

4. What can you conclude about the condition of the secondary wires?

____

____

Spark Plugs

1. Remove the engine's spark plugs. Place them on a bench arranged according to the cylinder number. Carefully examine the electrodes and porcelain of each plug and describe each here.

____

____

____

____

2. What should a normal spark plug look like?

____

____

3. Measure the gap of each spark plug and record your findings. NOTE: The gap of iridium spark plugs should not be measured. This can cause damage to the delicate electrode.

____

____

____

____

4. What is the specified gap? ______ inches

5. What is the apparent condition of the plugs?

____

____

6. Measure the resistance between the spark plug terminal and the center electrode. What was your measurement? ______ ohms

7. Why would a spark plug have resistance?

____

____

8. Summarize your findings.

____

____

____

9. Reinstall the plugs and torque them to specifications. What are the specs?

____

10. Reinstall the secondary wires. Task Completed ☐

Problems Encountered

____

____

____

Instructor's Comments

____

____

____

ENGINE PERFORMANCE JOB SHEET 31

Diagnosing EFI Systems

Name ______________________ Station ______________ Date ______________

NATEF Correlation

This Job Sheet addresses the following **MAST** task:

D.1. Diagnose (troubleshoot) hot or cold no-starting, hard starting, poor drivability, incorrect idle speed, poor idle, flooding, hesitation, surging, engine misfire, power loss, stalling, poor mileage, dieseling, and emissions problems; determine necessary action.

Objective

Upon completion of this job sheet, you will be able to diagnose the air/fuel system to identify the cause of hot or cold no-starting, hard starting, poor drivability, incorrect idle speed, poor idle, flooding, hesitation, surging, engine misfire, power loss, stalling, poor mileage, dieseling, and emissions problems and determine what services will correct the concern.

Tools and Materials

Service information

Lab scope

Noid light

Digital multimeter

Protective Clothing

Goggles or safety glasses with side shields

Describe the vehicle being worked on:

Year ________ Make ____________________ Model ____________________

VIN ____________________ Engine type and size ____________________

PROCEDURE

NOTE: *There are many things that can cause drivability problems. The items included in this job sheet are systems that most often cause concerns. Before proceeding to check the fuel system, check the condition of the engine and the ignition system.*

Inspecting the Fuel System

1. The fuel tank should be inspected for leaks; road damage; corrosion and rust on metal tanks; loose, damaged, or defective seams; loose mounting bolts; and damaged mounting straps. Summarize your findings.

____

____

____

2. Fuel lines should be inspected for holes, cracks, leaks, kinks, or dents. Summarize your findings.

____

____

3. Steel tubing should be inspected for leaks, kinks, and deformation. This tubing should also be checked for loose connections and proper clamping to the chassis. If the fuel tubing threaded connections are loose, they must be tightened to the specified torque. Some threaded fuel line fittings contain an O-ring. If the fitting leaks, the problem may be the O-ring. Summarize your findings.

____

____

____

4. Nylon fuel pipes should be inspected for leaks, nicks, scratches and cuts, kinks, melting, and loose fittings. If these fuel pipes are kinked or damaged in any way, they must be replaced. Nylon fuel pipes must be secured to the chassis at regular intervals to prevent fuel pipe wear and vibration. Summarize your findings.

____

____

____

5. The rubber fuel hose should be inspected for leaks, cracks, cuts, kinks, oil soaking, and soft spots or deterioration. If any of these conditions are found, the hose should be replaced. Summarize your findings.

____

____

____

6. Before proceeding with specific fuel injection checks and electronic control testing, be certain the battery is in good condition, fully charged, with clean terminals and connections. Also verify that the charging and starting systems are operating properly. Make sure all fuses and fusible links are intact and all wiring harnesses are properly routed with connections free of corrosion and tightly attached. What did you find?

____

____

7. Make sure all vacuum lines are in sound condition, properly routed, and tightly attached. What did you find?

____

____

8. Check the PCV system to make sure it is working properly. Also make sure all emission control systems are in place, hooked up, and operating properly. What did you find?

____

____

9. Check the coolant level and the operation of the cooling system and fans. What did you find?

____

____

10. Verify that the ignition system is functioning properly and that the idle speed is correct. What did you find?

____

____

11. Make sure the gasoline in the tank is of good quality and has not been substantially cut with alcohol or contaminated with water. What did you find?

____

____

12. Connect a scan tool to the vehicle and check for DTCs. Interpret the code and follow the diagnostic procedures given in the service information to identify the cause of the problem. What codes did you retrieve?

____

____

13. Remember that trouble codes only indicate the particular circuit in which a problem has been detected. They do not pinpoint individual problems or components. If a code suggests a component is faulty, that component needs to be thoroughly checked. The following procedures are divided by major components of the EFI system. If you suspect that one or more of these are faulty, move to that section of the job sheet. What components do you need to test to complete your diagnosis of the system?

____

____

Diagnosing Fuel Control Problems

1. The EFI system must have good communications with the vehicle network. Were there any communication DTCs? If so, what should you do next?

____

____

2. The STFT and LTFT should be looked at before beginning any other diagnosis of a drivability problem. Fuel trim allows you to look at what the PCM is doing to control fuel delivery. Was there a fuel trim DTC?

____

____

3. Fuel trim should be observed whether or not a DTC was set. Start the engine and determine if the system is in closed or open loop?

____

____

4. In regards to STFT and LTFT, how should they react when the system is in open loop?

____

____

5. With the system in closed loop, observe and record the STFT and the LTFT. What do they indicate?

____

____

6. A change in fuel trim is presented as a percentage. What is represented by a fuel trim number that has a minus sign before it?

____

____

7. On the scan tool, observe the oxygen sensor and the STFT. Does the STFT toggle with the oxygen sensor's output?

____

____

8. What is represented by the observed LTFT?

____

____

9. If there is a constant rich condition, how will the LTFT respond?

____

____

10. With the engine running, pull off a large vacuum hose and watch the STFT and LTFT values. What did they do and what does this indicate?

____

____

11. Based on these checks, what do you conclude about the PCM's ability to control the air/fuel mixture?

____

____

____

Oxygen Sensor Diagnosis

1. Make sure the engine is at a normal operating temperature before testing the oxygen (O_2) sensor. What is the normal temperature?

____

2. Refer to the vehicle's wiring diagram to identify the terminals at the sensor. Heated oxygen sensors have four wires connected to them. Two are for the heater and the other two are for the sensor. Identify the terminals of the O_2 sensors on this vehicle. Describe them here.

____

____

3. An O_2 sensor can be checked with a voltmeter. Connect it between the O_2 sensor wire and ground. With the engine running, the sensor's voltage should be cycling from 0 and 1 volt. If the voltage is continually high, the air–fuel ratio may be rich or the sensor may be contaminated. If the O_2 sensor voltage is continually low, the air–fuel ratio may be lean, the sensor may be defective, or the wire between the sensor and the computer may have a high-resistance problem. If the O_2 sensor voltage signal remains in a mid-range position, the computer may be in open loop or the sensor may be defective. Summarize the results of this check.

____

____

____

4. The activity of an O_2 sensor is best monitored with a lab scope. The switching of the sensor should be seen as the sensor signal goes from lean to rich to lean continuously. Summarize the results of this check.

____

____

____

5. The activity of the sensor can also be monitored on a scanner. By watching the scanner while the engine is running, the O_2 voltage should move to nearly 1 volt, and then drop back to close to zero volts. Immediately after it drops, the voltage signal should move back up. This immediate cycling is an important function of an O_2 sensor. If the response is slow, the sensor is lazy and should be replaced. With the engine at about 2,500 rpm, the O_2 sensor should cycle from high to low 10 to 40 times in 10 seconds.

When testing the O_2 sensor, make sure the sensor is heated and the system is in a closed loop. Summarize the results of this check.

____

____

____

Air/Fuel Ratio Sensor

Many vehicles are equipped with air–fuel ratio (A/F) sensors, also called wideband O_2 sensors, that operate differently from a conventional oxygen sensor. The A/F ratio sensor has the ability to better detect air–fuel ratios away from the normal 14.7 to 1, giving the ECM a more accurate fuel delivery than before. The air–fuel ratio sensor may still be called an oxygen sensor in the service information. Typically, an A/F sensor will have around 7 wires in the connector, instead of the 4 wires in the conventional heated O_2 sensor. The A/F ratio sensor operates on a very small current flow, which changes direction to indicate a rich or lean condition. When the mixture is rich, the current flow is in a negative direction, when lean in a positive direction, and at stoichiometry, there is no current flow.

1. Does your vehicle use an A/F ratio sensor? ____________________
2. If so, what is the proper way to diagnose the sensor? ______________
3. Is it advisable to take direct measurements from the sensor wiring?

   ____

4. What is the procedure to diagnose the sensor properly?

   ____

   ____

   ____

Air Induction System Checks

1. Check the air control system for cracks and deteriorated ductwork. Also make sure all the induction hose clamps are tight and properly sealed. Look for possible air leaks in the crankcase—for example, around the dipstick tube and oil filter cap. Summarize the results of this check.

   ____

   ____

   ____

2. Check the vacuum lines and components. Summarize the results of this check.

   ____

   ____

   ____

Airflow Sensors

1. If the MAF sensor is in the air duct or attached to the air cleaner, inspect the duct for tears that could open up on acceleration. What did you find?

____

____

____

2. Check the wiring connection to the MAF sensor. Is the connection clean and tight? Do the connectors look like they are making good contact?

____

____

____

3. Use a scan tool to check the operation of a mass airflow meter. Record the grams/per/second readings at the following throttle opening.

   Idle ____

   1,000 rpm ____

   2,000 rpm ____

Throttle Body Diagnosis (Cable-Operated Throttle)

1. Check the inside of the throttle body assembly for accumulations of dirt, carbon, or other substances, especially on the backside of the throttle. Summarize the results of this check.

____

____

____

2. If necessary, clean the area behind the throttle plate. This accumulation can cause stalling on deceleration. Was this necessary?

____

Vehicles with Throttle Actuator Control (TAC)

1. Remove the air duct from the throttle body and check the service information for determining the position of the throttle plate. Now with the key off, note the position of the throttle plate.

____

____

2. Have an assistant turn the key on (engine off) and push the accelerator pedal slowly open and closed while you observe the throttle. Does the throttle respond to the accelerator pedal correctly?

____

____

3. Connect a scan tool to the vehicle and command the throttle open and closed with the scan tool. Does the throttle respond correctly to the commands?

____

____

4. Watching the scan tool, observe the accelerator position sensors and the TP sensors. Do they respond as expected (looking at the service information)?

____

5. What conclusion can you draw from your observations?

____

____

Fuel System Checks

1. Check fuel pressure and fuel volume. Compare your measurements with the specifications and state your conclusions about the fuel system.

____

____

2. Is the system you have a returnless system, or does it have a return line?

____

____

____

3. If the fuel pressure slowly bleeds down, there may be a leak in the fuel pressure regulator, fuel pump check valve, or the injectors themselves. Remember, hard starting is a common symptom of system leaks. Summarize the results of this check.

____

____

____

Injector Checks

1. Warm up a gas analyzer and set the meters to ambient air. What model exhaust analyzer will you be using and how long does the manufacturer recommend that it be run to warm up? Also, describe the other procedures required to prepare the analyzer for testing.

___

___

___

2. With the engine warmed up, but not running, remove the air duct from the airflow sensor. Then, insert the gas analyzer's probe into the intake plenum area. Look at the HC readings on the analyzer. They should be low and drop as time passes. If an injector is leaking, the HC reading will be high and will not drop. Summarize the results of this check.

___

___

___

3. Another cause of a rich mixture is a leaking fuel pressure regulator. After the engine has been run, disconnect the vacuum line to the fuel pressure regulator (if applicable). If there is any sign of fuel inside the hose or if fuel comes out of the hose, the regulator's diaphragm is leaking. The regulator can also be tested with a hand-operated vacuum pump. Apply 5 in. Hg to the regulator. A good regulator diaphragm will hold that vacuum. Summarize the results of this check.

___

___

___

4. Check to see if the injectors are receiving a signal from the PCM to fire. Remove the injector's electrical connector and check for voltage at the injector using a high-impedance test light or a noid light. While cranking the engine, the light should flash if the computer is cycling the injector on and off. If the light is not flashing, the computer or connecting wires are defective. Summarize the results of this check.

___

___

___

5. An ohmmeter can be used to test an injector. Connect the ohmmeter across the injector terminals after the wires to the injector have been disconnected. If the meter reading is infinite, the injector winding is open. If the meter shows more resistance than the specifications call for, there is high resistance in the winding. A reading that is lower than the specifications

indicates that the winding is shorted. Compare your readings with the specifications and explain your conclusions about the injectors.

____

____

____

6. Connect the lab scope's positive lead to the injector supply wire and the scope's negative lead to an engine ground. Set the scope to read 12 volts, and then set the sweep and trigger to allow you to clearly see the "on" signal on the left and the "off" signal on the right. Make sure the entire waveform is clearly seen. Also, remember that the setting may need to be changed as engine speed increases or decreases. Observe and describe the pattern of the fuel injectors.

____

____

____

7. When checking the injectors with a lab scope, make sure the injectors are firing at the correct time. To do this, use a dual trace scope and monitor the ignition reference signal and a fuel injector signal at the same time. The two signals should have some sort of rhythm between them. This rhythm is dependent upon several things; however, it doesn't matter what the rhythm is, it only matters that the rhythm is constant. If the injector's waveform is fine but the rhythm varies, the ignition reference sensor circuit is faulty and is not allowing the injector to fire at the correct time. Summarize the results of this check.

____

____

____

Problems Encountered

____

____

____

Instructor's Comments

____

____

____

ENGINE PERFORMANCE JOB SHEET 32

Checking Fuel for Contaminants

Name ______________________ Station ______________ Date ______________

NATEF Correlation

This Job Sheet addresses the following **AST** task:

D.1. Check fuel for contaminants; determine necessary action.

This Job Sheet addresses the following **MAST** task:

D.2. Check fuel for contaminants; determine necessary action.

Objective

Upon completion of this job sheet, you will be able to check the fuel on a vehicle for contaminants and quality.

Tools and Materials

Calibrated cylinder

Protective Clothing

Goggles or safety glasses with side shields

Describe the vehicle being worked on:

Year ________ Make ____________________ Model ____________________

VIN ____________________ Engine type and size ____________________

PROCEDURE

1. Pump gasoline may contain a small amount of alcohol, normally up to 10%, unless the vehicle is equipped to burn E-85. However, for most vehicles, if the amount is greater than that, problems may result such as fuel system corrosion, fuel filter plugging, deterioration of fuel system components, and a lean air–fuel ratio. These fuel system problems caused by excessive alcohol in the fuel may result in drivability complaints such as lack of power, acceleration stumbles, engine stalling, and no-start. If the correct amount of fuel is being delivered to the engine and there is evidence of a lean mixture, check for vacuum leaks in the intake, and then check the gasoline's alcohol content. Did you find any vacuum leaks in the system? Where?

   ____

   ____

2. Obtain a 100-milliliter (mL) cylinder graduated in 1-mL divisions. Task Completed ☐
3. Fill the cylinder to the 90-mL mark with gasoline. Task Completed ☐

4. Add 10 mL of water to the cylinder so it is filled to the 100-mL mark. Task Completed ☐
5. Install a stopper in the cylinder, and shake it vigorously for 10 to 15 seconds. Task Completed ☐
6. Carefully loosen the stopper to relieve any pressure. Task Completed ☐
7. Install the stopper and shake vigorously for another 10 to 15 seconds. Task Completed ☐
8. Carefully loosen the stopper to relieve any pressure. Task Completed ☐
9. Place the cylinder on a level surface for 5 minutes to allow liquid separation. Task Completed ☐
10. Any alcohol in the fuel is absorbed by the water and settles to the bottom. If the water content in the bottom of the cylinder exceeds 10 mL, there is alcohol in the fuel. Task Completed ☐
11. How much alcohol was measured?

    ____

12. Were there any particles that settled to the bottom? Describe them.

    ____

    ____

13. What are your service recommendations?

    ____

    ____

Problems Encountered

____

____

____

Instructor's Comments

____

____

____

ENGINE PERFORMANCE JOB SHEET 33

Checking the Fuel Delivery System

Name ______________________________ Station __________________ Date __________________

NATEF Correlation

This Job Sheet addresses the following **AST** task:

D.2. Inspect and test fuel pumps and pump control systems for pressure, regulation, and volume; perform necessary action.

This Job Sheet addresses the following **MAST** task:

D.3. Inspect and test fuel pumps and pump control systems for pressure, regulation, and volume; perform necessary action.

Objective

Upon completion of this job sheet, you will be able to remove, inspect, and test the vacuum and electrical components and circuits of a fuel delivery system, check the system for leaks, inspect and test electric fuel pumps, and inspect and test fuel pressure regulation systems.

Tools and Materials

Clean shop rags
Approved gasoline container
Pressure gauge with adapters
Graduated container
Fuel hose approximately 2 feet in length
Hand-operated vacuum pump
DMM
Service information

Protective Clothing

Goggles or safety glasses with side shields

Describe the vehicle being worked on:

Year __________ Make __________________________ Model __________________________

VIN __________________________ Engine type and size __________________________

Describe general operating condition:

____

____

PROCEDURE

1. Connect a scan tool and record all DTCs.

____

____

2. Were any of the DTCs related to the fuel system? What do they indicate?

____

____

NOTE: *Since electronic fuel injection systems have a residual fuel pressure, this pressure must be relieved before disconnecting any fuel system component. Failure to relieve the fuel pressure on electronic fuel injection (EFI) systems prior to fuel system service may result in gasoline spills, serious personal injury, and expensive property damage. Direct fuel injection systems have extremely high pressure in the fuel rail, as well as pressure in the supply section of the fuel system.*

3. Loosen the fuel tank filler cap to relief any fuel tank vapor pressure. Did you feel any pressure?

____

____

4. Wrap a shop towel around the fuel pressure test port on the fuel rail and remove the dust cap from this valve. Install the bleed hose on the gauge in an approved gasoline container and open the gauge bleed valve to relieve fuel pressure from the system into the gasoline container. Be sure all the fuel in the bleed hose is drained into the gasoline container. How much fuel was released?

____

____

5. Dispose of the fuel in the proper way. What did you do with it?

____

____

Inspecting and Testing the Fuel Pump and Its Circuit

1. Carefully inspect the fuel lines and connectors. Did you find anything abnormal?

____

____

2. Carefully inspect the fuel rail and injectors for signs of leaks. Record your findings.

____

____

3. Look up the specifications for the fuel pump. The fuel pump pressure specifications are: __________ psi

4. Connect the fuel pressure tester to the Schrader valve on the fuel rail. Turn the ignition on and observe the fuel pressure readings. Your readings are: __________ psi.

5. Compare the readings to specifications. What is indicated by the readings?

____

____

If your vehicle has a return line, complete numbers 6–8.

6. With the engine running, disconnect the vacuum hose at the pressure regulator. Observe the fuel pressure. Did the pressure respond as you thought it should? Why or why not?

____

____

7. Connect a hand-operated vacuum pump to the fuel pressure regulator. Create a vacuum at the pressure regulator and observe the fuel pressure. Then, relieve the vacuum. Describe what happened.

____

____

8. If the fuel system has a fuel return line at the fuel rail, pinch it shut with pinch-off pliers. NOTE: Never pinch off a nylon fuel line. If the line is nylon or steel, special adaptors may have to be installed to perform this test. (See the service information.) Observe the fuel pressure gauge. What happened and why?

____

If the system does not have a return line, do task numbers 9–11.

9. Attach a fuel pressure gauge and allow the engine to run long enough for the fuel pressure to stabilize. Turn the ignition off and watch the fuel pressure gauge. Does it fall to less than half of its original value (or the manufacturer's specified value)?

____

____

____

10. If the fuel pressure dropped more than the specified value, run the fuel pump to allow the fuel pressure to build back to normal. Clamp the fuel feed line at the fuel rail (use a fuel line adaptor if the line is steel or nylon). Recheck the rate of fuel drop. Does the fuel pressure still drop?

____

____

11. If the pressure did not drop with the fuel line blocked, the problem is the fuel pump or comes from the fuel lines to the rail. Inspect the lines carefully for leaks. If no leaks are found in the lines, what is most likely the source of the loss of pressure?

____

____

____

12. If the pressure drops with the fuel line blocked, what is the cause of the drop in fuel pressure?

____

____

____

13. Turn off the ignition. Relieve the fuel system pressure. Disconnect the fuel return hose at the regulator or fuel rail if it's a returnless system. Connect a hose to that port and insert the open end of the hose into a graduated container suited for gasoline. It is best if the container can hold more than one quart (1 liter). How much can the container hold, and what increments are evident on the container?

____

____

14. Turn the ignition on and allow the fuel pump to pump fuel into the container for 7 seconds. Then, turn off the ignition. How much fuel is in the container?

____

____

15. How much should the pump be able to supply according to specifications?

____

____

16. What are your service recommendations?

____

____

Problems Encountered

____

____

____

Instructor's Comments

____

____

____

ENGINE PERFORMANCE JOB SHEET 34

Replacing a Fuel Filter

Name ______________________ Station ______________ Date ______________

NATEF Correlation

This Job Sheet addresses the following **MLR** task:

C.1. Replace fuel filter(s).

This Job Sheet addresses the following **AST** task:

D.3. Replace fuel filter(s).

This Job Sheet addresses the following **MAST** task:

D.4. Replace fuel filter(s).

Objective

Upon completion of this job sheet, you will be able replace a fuel filter.

Tools and Materials

Basic hand tools
Shop towels
Fuel pressure gauge set
Approved gasoline container

Protective Clothing

Goggles or safety glasses with side shields

Describe the vehicle being worked on:

Year ________ Make ____________________ Model ____________________

VIN ____________________ Engine type and size ____________________

Describe general operating condition:

____

____

____

PROCEDURE

1. Disconnect the negative cable at the battery. Task Completed ☐
2. Loosen the fuel tank filler cap to relieve any fuel tank vapor pressure. Task Completed ☐
3. Wrap a shop towel around the Schrader valve on the fuel rail and remove the dust cap from the valve. Task Completed ☐
4. Connect the fuel pressure gauge to the Schrader valve. Task Completed ☐

5. Install the free end of the gauge bleed hose into an approved gasoline container and slowly open the gauge bleed valve to relieve the fuel pressure. Task Completed ☐

6. Place the vehicle on the hoist and position the lift arms according to the manufacturer's recommendations. Then, raise the vehicle. Task Completed ☐

7. Locate the fuel filter. Describe its location.

____

____

8. Flush the fuel filter line connectors with water and use compressed air to blow debris off and away from the connectors. Task Completed ☐

9. Follow the recommended procedures for disconnecting the fuel inlet connector. What is the procedure and what special tools are required?

____

____

10. Follow the recommended procedures for disconnecting the fuel outlet connector. What is the procedure and what special tools are required?

____

____

11. Remove the fuel filter. Task Completed ☐

12. Before installing a new filter, clean the ends of the new filter. Task Completed ☐

13. Apply a light coating of clean oil onto the fittings. Task Completed ☐

14. Check the quick connectors to be sure the large collar on each connector has rotated back to its original position. Task Completed ☐

15. Install the new filter, making sure it is facing the correct direction. Task Completed ☐

16. Press the connectors and lines firmly onto the filter. Task Completed ☐

17. Tighten the filter retaining bolts and brackets. Task Completed ☐

18. Start the engine and check for leaks. Task Completed ☐

Problems Encountered

____

____

____

Instructor's Comments

____

____

____

ENGINE PERFORMANCE JOB SHEET 35

Servicing a Throttle Body / Replacing an Air Filter

Name ______________________ Station ______________ Date ______________

NATEF Correlation

This Job Sheet addresses the following **MLR** task:

C.2. Inspect, service, or replace air filters, filter housings, and intake duct work.

This Job Sheet addresses the following **AST** tasks:

D.4. Inspect, service, or replace air filters, filter housings, and intake duct work.

D.5. Inspect throttle body, air induction system, intake manifold and gaskets for vacuum leaks and/or unmetered air.

This Job Sheet addresses the following **MAST** tasks:

D.5. Inspect, service, or replace air filters, filter housings, and intake duct work.

D.6. Inspect throttle body, air induction system, intake manifold and gaskets for vacuum leaks and/or unmetered air.

Objective

Upon completion of this job sheet, you will be able to inspect the throttle body mounting plates, air induction and filtration systems, intake manifolds, and gaskets. You will also be able to remove, service, and install a throttle body.

Tools and Materials

12-volt power supply or memory keeper
Throttle body cleaner
Compressed air
OSHA-approved air nozzle
Hand tools
Service information

Protective Clothing

Goggles or safety glasses with side shields

Describe the vehicle being worked on:

Year __________ Make ____________________ Model ____________________

VIN ____________________ Engine type and size ____________________

Describe general operating condition:

____

____

PROCEDURE

Visually Inspect an EFI System

1. Is the battery in good condition, fully charged, with clean terminals and connections? ☐ Yes ☐ No
2. Do the charging and starting systems operate properly? ☐ Yes ☐ No
3. Are all fuses and fusible links intact? ☐ Yes ☐ No
4. Are all wiring harnesses properly routed, with connections free of corrosion and tightly attached? ☐ Yes ☐ No
5. Are all vacuum lines in sound condition, properly routed, and tightly attached? ☐ Yes ☐ No
6. Is the PCV system working properly and maintaining a sealed crankcase?
7. Are all emission control systems in place, hooked up, and operating properly? ☐ Yes ☐ No
8. Is the level and condition of the coolant/antifreeze good and is the thermostat opening at the proper temperature? ☐ Yes ☐ No
9. Are the secondary spark delivery components in good shape, with no signs of crossfiring, carbon tracking, corrosion, or wear? ☐ Yes ☐ No
10. Is the idle speed set to specifications? ☐ Yes ☐ No
11. Does the air intake duct work have cracks or tears? ☐ Yes ☐ No
12. Are all of the induction hose clamps tight and properly sealed? ☐ Yes ☐ No
13. Are there any other possible air leaks in the crankcase? ☐ Yes ☐ No
14. Are the electrical connections at the mass airflow sensor or manifold pressure sensor good? ☐ Yes ☐ No
15. Is there carbon buildup inside the throttle bore and on the throttle plate? ☐ Yes ☐ No
16. Is there a clicking noise at each injector while the engine is running? ☐ Yes ☐ No
17. Check the condition of the air filter. Does it show signs of needing replacement? How did you make your determination?

    ____

18. Before replacing the air filter (if necessary), make sure the air box is free of debris. ☐ Yes ☐ No

Servicing a Throttle Body

NOTE: *Whenever it is necessary to remove the throttle body assembly for replacement or cleaning, make sure you follow the procedures outlined by the manufacturer. Some throttle bodies have a special coating that should not be cleaned with solvents. Check the service information before starting any work.*

1. Disconnect the negative battery cable. Task Completed ☐
2. Remove the air ducting from the throttle body. Task Completed ☐

3. Unbolt and remove the throttle body. Task Completed ☐

4. Once the assembly has been removed, remove all nonmetallic parts such as the TP sensor, IAC valve, throttle opener, and the throttle body gasket from the throttle body. Never soak any electrical components in solvent. Task Completed ☐

5. It is now safe to clean the throttle body assembly in the recommended cleaner and blow dry it with compressed air. Make sure you blow out all passages in the throttle body assembly. Task Completed ☐

6. Before reinstalling the throttle body assembly, check to make sure all metal mating surfaces are clean and free from metal burrs and scratches. Task Completed ☐

7. With new gaskets and seals, install the assembly. Tighten all fasteners to the recommended torque. Task Completed ☐

8. After everything that was disconnected is reconnected, reconnect the negative battery cable. Task Completed ☐

9. After all the parts are reconnected and installed, adjust the throttle linkage according to the manufacturer's recommendations. Task Completed ☐

10. Relearn idle speeds as necessary. Task Completed ☐

Problems Encountered

____

____

____

Instructor's Comments

____

____

____

ENGINE PERFORMANCE JOB SHEET 36

Inspect, Clean, and Test Fuel Injectors

Name ______________________ Station ______________ Date ______________

NATEF Correlation

This Job Sheet addresses the following **AST** task:

D.6. Inspect and test fuel injectors.

This Job Sheet addresses the following **MAST** task:

D.7. Inspect and test fuel injectors.

Objective

Upon completion of this job sheet, you will be able to properly inspect, test, and clean fuel injectors.

Tools and Materials

Fuel pressure gauge	Lab scope
Test spark plug	Noid light
Digital multimeter	Service information
Injector cleaner	Injector tester

Protective Clothing

Goggles or safety glasses with side shields

Describe the vehicle being worked on:

Year ________ Make ____________________ Model ____________________

VIN ____________________ Engine type and size ____________________

Describe general operating condition:

____

____

PROCEDURE

1. Connect a scan tool to the vehicle and check for DTCs. Interpret the code and follow the diagnostic procedures given in the service information to identify the cause of the problem. What codes did you retrieve?

____

____

2. Remember that trouble codes only indicate the particular circuit in which a problem has been detected. They do not pinpoint individual problems or components. If a code suggests a component is faulty, that component needs to be thoroughly checked. The following procedures are divided by major components of the EFI system. If you suspect that one or more of these are faulty, move to that section of the job sheet. What components do you need to test to complete your diagnosis of the system?

____

____

3. The EFI system must have good communications with the vehicle network. Were there any communication DTCs? If so, what should you do next?

____

____

4. The STFT and LTFT should be looked at before beginning any other diagnosis of a drivability problem. Fuel trim allows you to look at what the PCM is doing to control fuel delivery. Was there a fuel trim DTC?

____

____

5. Fuel trim should be observed whether or not a DTC was set. Start the engine and watch the STFT and LTFT.

Task Completed ☐

6. A change in fuel trim is presented as a percentage. What is represented by a fuel trim number that has a minus sign before it?

____

____

7. On the scan tool, observe the oxygen sensor and the STFT. Does the STFT toggle with the oxygen sensor's output?

____

____

8. What is represented by the observed LTFT?

____

____

9. If there is a constant rich condition, how will the LTFT respond?

____

____

10. With the engine running, pull off a large vacuum hose and watch the STFT and LTFT values. What did they do and what does this indicate?

____

____

11. Based on these checks, what do you conclude about the PCM's ability to control the air–fuel mixture?

____

____

12. Check the fuel pressure at the fuel rail. A pressurized fuel supply must be delivered to the properly operating injectors. If there is no pressure, refer to the service information to determine if there is a switch or relay that will shut off the fuel pump if a collision or large jar is detected by the switch. Summarize your findings.

____

____

13. Check for voltage signals to the injectors; the injectors must receive a trigger signal to inject the fuel. Summarize your findings.

____

____

14. Check the inside of the throttle body assembly for accumulations of dirt, carbon, or other substances. Summarize the results of this check.

____

____

15. Check to see if the injectors are receiving a signal from the PCM to fire. Remove the injector's electrical connector and check for voltage at the injector using a high-impedance test light or a noid light. While cranking the engine, the light should flash if the computer is cycling the injector on and off. If the light is not flashing, the computer or connecting wires are defective. Summarize the results of this check.

____

____

Testing Injectors

The best way to test fuel injectors is through an injector balance test. The injector balance test measures the amount of fuel dropped by the injector during operation. The balance tester can be used to activate the injector much like the ECM. The technician can watch the amount of fuel dropped by each injector by observing the fuel pressure gauge.

1. Begin by installing a fuel pressure gauge on the fuel rail. Activate the fuel pump at least twice for a few seconds. Allow the fuel pressure reading to stabilize, and then record the reading in the following chart. Task Completed ☐

2. Connect the fuel injector tester to the vehicle battery. Connect the tester to a fuel injector. Task Completed ☐

3. Activate the fuel tester. (Most have a setting to activate the injection pulse 10 or 100 times to check the mechanical operation of the injector.) Allow the fuel pressure reading to stabilize, and then record this second reading for each injector in the chart. Task Completed ☐

4. Compute the pressure drop for each injector and record those values in the chart. Task Completed ☐

5. Compare your readings to the specifications. Are there any that are out of limits? What are your recommendations?

____

____

Injector	*Initial pressure reading*	*Pressure reading after activation*	*Pressure drop*
#1			
#2			
#3			
#4			
#5			
#6			
#7			
#8			

Lab Scope Testing of Injector Waveforms

1. Connect the lab scope's positive lead to the injector supply wire and the scope's negative lead to an engine ground. Set the scope to read 12 volts, and then set the sweep and trigger to allow you to clearly see the "on" signal on the left and the "off" signal on the right. Make sure the entire waveform is clearly seen. Also remember that the setting may need to be changed as the engine speed increases or decreases. Observe and describe the pattern of the fuel injectors.

____

____

2. When checking the injectors with a lab scope, make sure the injectors are firing at the correct time. To do this, use a dual trace scope and monitor the ignition reference signal and a fuel injector signal at the same time. The two signals should have some sort of rhythm between them. This rhythm is dependent upon several things; however, it doesn't matter what the rhythm is, it only matters that the rhythm is constant. If the injector's

waveform is fine but the rhythm varies, the ignition reference sensor circuit is faulty and is not allowing the injector to fire at the correct time. Summarize the results of this check.

____

____

Cleaning Injectors

If an injector fails the balance test, or the LTFT fuel trim is trending lean with no vacuum leaks, an injector cleaning might be in order.

Automotive parts stores usually sell pressurized containers of injector cleaner with a hose for Schrader valve attachment. During the cleaning process, the engine is operated on the pressurized container of unleaded fuel and injector cleaner. Fuel pump operation must be stopped to prevent the pump from forcing fuel up to the fuel rail. The fuel return line is to be plugged to prevent the solution in the cleaning container from flowing through the return line into the fuel tank. Always follow the instructions and precautions on the particular product you are using.

1. Disconnect the wires from the in-tank fuel pump or the fuel pump relay to disable the fuel pump. If you disconnect the fuel pump relay on General Motors products, the oil pressure switch in the fuel pump circuit must also be disconnected to prevent current flow through this switch to the fuel pump. Task Completed ☐
2. Plug the fuel return line from the fuel rail to the tank. Task Completed ☐
3. Connect a can of injector cleaner to the Schrader valve on the fuel rail, and run the engine for about 20 minutes on the injector solution. Task Completed ☐

Replacing an Injector

CAUTION: *Often an individual injector needs to be replaced. Random disassembly of the components and improper procedures can result in damage to one of the various systems located near the injectors.*

1. The injectors are normally attached directly to a fuel rail and inserted into the intake manifold or cylinder head. They must be positively sealed because high-pressure fuel leaks can cause a serious safety hazard. What components are located near the fuel injectors?

____

____

2. Prior to loosening any fitting in the fuel system, the fuel pump fuse should be removed. Where is this fuse located? As an extra precaution, disconnect the negative cable at the battery.

____

____

3. To remove an injector, the fuel rail must be able to move away from the engine. The rail holding brackets should be unbolted and the vacuum line

to the pressure regulator disconnected. Disconnect the wiring harness to the injectors. What must be done to disconnect the harness?

____

____

4. The injectors are held to the fuel rail by a clip that fits over the top of the injector. An O-ring at the top and at the bottom of the injector seals the injector. Pull up on the fuel rail assembly. The bottom of the injectors will pull out of the manifold while the tops are secured to the rail by clips. Remove the clip from the top of the injector and remove the injector unit. Did you have any problems doing this?

____

____

5. Install new O-rings onto the new injector. Be careful not to damage the seals while installing them and make sure they are in their proper locations. Install the injector into the fuel rail and set the rail assembly into place. Tighten the fuel rail hold-down bolts according to the manufacturer's specifications. What are the specifications?

____

____

6. Reconnect all parts that were disconnected. Install the fuel pump fuse and reconnect the battery. Turn the ignition switch to the run position and check the entire system for leaks. Did you find any?

____

____

7. After a visual inspection has been completed, conduct a fuel pressure test on the system. What do the results of the pressure test indicate?

____

____

Problems Encountered

____

____

____

Instructor's Comments

____

____

____

ENGINE PERFORMANCE JOB SHEET 37

Verify Idle Control Operation

Name ______________________ Station ______________ Date ______________

NATEF Correlation

This Job Sheet addresses the following **AST** task:

D.7. Verify idle control operation.

This Job Sheet addresses the following **MAST** task:

D.8. Verify idle control operation.

Objective

Upon completion of this job sheet, you will be able to check the engine idle speed on a late-model vehicle.

Tools and Materials

Service information

Scan tool

Protective Clothing

Goggles or safety glasses with side shields

Describe the vehicle being worked on:

Year ________ Make ____________________ Model ____________________

VIN ____________________ Engine type and size ____________________

Describe general operating condition:

____

____

PROCEDURE

1. Before checking the idle speed, check that the MIL is not lit. If it is, diagnose the cause. What did you find?

   ____

   ____

2. Visually inspect the air cleaner assembly and PCV system for damage. Also, check to make sure there are no apparent ignition problems. What did you find?

   ____

   ____

3. Does the manufacturer call for an idle learn sequence? If so, briefly explain the process.

____

____

4. Start the engine. Hold the engine at 3,000 rpm without load (in park or neutral) until the radiator cooling fan comes on, then let it idle. Check the idle speed with the headlights, blower fan, radiator fan, and air conditioner off. What was the engine speed?

____

____

5. Let the engine idle for 1 minute with the heater fan and air conditioner off. Check the idle speed and compare it to specifications. Is the idle speed correct?

____

____

6. Let the engine idle for 1 minute with the headlights, blower fan, radiator fan, and air conditioner on. Check the idle speed and compare to specifications. Is the idle speed correct? What should be done to correct it?

____

____

7. Turn off the engine. If the idle speed was not correct, check the throttle cable for binding or excessive freeplay. What did you find?

____

____

8. If the throttle cable needs adjustment, describe the procedure for doing this.

____

____

9. With the cable properly adjusted, make sure the throttle plate opens fully when you push the accelerator pedal to the floor. Also, make sure it returns to the idle position when the accelerator pedal is released. If the vehicle is equipped with electronic throttle control, watch the throttle while an assistant moves the accelerator pedal. What did you find?

____

____

10. If no problems were evident, check the intake manifold and throttle body for vacuum leaks. What did you find?

____

____

Problems Encountered

____

____

____

Instructor's Comments

____

____

____

ENGINE PERFORMANCE JOB SHEET 38

Inspect Exhaust System

Name ______________________________ Station __________________ Date __________________

NATEF Correlation

This Job Sheet addresses the following **MLR** tasks:

C.3. Inspect integrity of the exhaust manifold, exhaust pipes, muffler(s), catalytic converter(s), resonator(s), tail pipe(s), and heat shields; determine necessary action.

C.4. Inspect condition of exhaust system hangers, brackets, clamps, and heat shields; repair or replace as needed.

This Job Sheet addresses the following **AST** tasks:

D.8. Inspect integrity of the exhaust manifold, exhaust pipes, muffler(s), catalytic converter(s), resonator(s), tail pipe(s), and heat shields; determine necessary action.

D.9. Inspect condition of exhaust system hangers, brackets, clamps, and heat shields; repair or replace as needed.

This Job Sheet addresses the following **MAST** tasks:

D.9. Inspect integrity of the exhaust manifold, exhaust pipes, muffler(s), catalytic converter(s), resonator(s), tail pipe(s), and heat shields; determine necessary action.

D.10. Inspect condition of exhaust system hangers, brackets, clamps, and heat shields; repair or replace as needed.

Objective

Upon completion of this job sheet, you will be able to properly inspect exhaust manifolds, exhaust pipes, mufflers, catalytic converters, resonators, tail pipes, and heat shields.

Tools and Materials

Flashlight or trouble light	Service information
Hammer or mallet	Tachometer
Lift	Vacuum gauge

Protective Clothing

Goggles or safety glasses with side shields

Describe the vehicle being worked on:

Year __________ Make ________________________ Model ________________________

VIN ________________________ Engine type and size ________________________

Describe general operating condition:

____

____

PROCEDURE

1. Before doing a visual inspection, listen closely for hissing or rumbling that may indicate the beginning of exhaust system failure. With the engine idling, slowly move along the entire system and listen for leaks. Task Completed ☐

   WARNING: *Be very careful. Remember that the exhaust system gets very hot. Do not get your face too close when listening for leaks.*

   a. Did you hear any indications of a leak? ☐ Yes ☐ No

   b. If so, where?

   ____

   ____

2. Safely raise the vehicle. Task Completed ☐

3. With a flashlight or trouble light, check for the following:
   - Holes and road damage
   - Discoloration and rust
   - Carbon smudges
   - Bulging muffler seams
   - Interfering rattle points
   - Torn or broken hangers and clamps
   - Missing or damaged heat shields

   a. Did you detect any of these problems? ☐ Yes ☐ No

   b. If so, where?

   ____

   ____

4. Sound out the system by gently tapping the pipes and muffler with a hammer or mallet. A good part will have a solid metallic sound. A weak or worn-out part will have a dull sound. Listen for falling rust particles on the inside of the muffler. Mufflers usually corrode from the inside out, so the damage may not be visible from the outside. Remember that some rust spots might be only surface rust.

   a. Did you find any weak or worn-out parts? ☐ Yes ☐ No

   b. If so, which ones?

   ____

   ____

5. Grab the tailpipe (when it is cool) and try to move it up and down and from side to side. There should be only slight movement in any direction. If the system feels wobbly or loose, check the clamps and hangers that fasten the tailpipe to the vehicle.

   a. Did you detect any problems with the clamps or hangers? ☐ Yes ☐ No

   b. If so, where?

   ____

   ____

6. Check all of the pipes for kinks and dents that might restrict the flow of exhaust gases.

   a. Did you find any kinks or dents? ☐ Yes ☐ No

   b. If so, where?

   ____

   ____

7. Take a close look at each connection, including the one between the exhaust manifold and exhaust pipe.

   a. Did you find any white powdery deposits? ☐ Yes ☐ No

   b. If so, try tightening the bolts or replacing the gasket at that connection. Task Completed ☐

   Not applicable ☐

   c. Check for loose connections at the muffler by pushing up on the muffler slightly. Task Completed ☐

   d. If loose, try tightening them. Task Completed ☐

   Not applicable ☐

8. If a visual inspection does not identify a partially restricted or blocked exhaust system, perform the following test.

   a. Attach a vacuum gauge to the intake manifold. Connect a tachometer. Start the engine and observe the vacuum gauge. It should indicate a vacuum of 16–20 in. of mercury. Does it? ☐ Yes ☐ No

   b. Increase the engine's speed to 2,000 rpm and observe the vacuum gauge. Vacuum will decrease when the speed is increased rapidly, but it should stabilize at 16–21 in. of mercury and remain constant. If the vacuum does not build up to at least the idle reading, the exhaust system is restricted or blocked. Is the system restricted or blocked? ☐ Yes ☐ No

9. Catalytic converters can overheat. Look for bluish or brownish discoloration of the outer stainless steel shell. Also, look for blistered or burned paint or undercoating above and near the converter.

   Are there any signs of overheating? ☐ Yes ☐ No

10. Look up and record the part numbers of any parts discovered to be defective in previous steps.

____

____

____

____

____

Problems Encountered

____

____

____

Instructor's Comments

____

____

____

ENGINE PERFORMANCE JOB SHEET 39

Test a Catalytic Converter for Efficiency

Name ______________________ Station ______________ Date ______________

NATEF Correlation

This Job Sheet addresses the following **AST** tasks:

D.10. Perform exhaust system back-pressure test; determine necessary action.

E.6. Inspect and test catalytic converter efficiency.

This Job Sheet addresses the following **MAST** tasks:

D.11. Perform exhaust system back-pressure test; determine necessary action.

E.9. Inspect and test catalytic converter efficiency.

Objective

Upon completion of this job sheet, you will be able to properly perform exhaust system back-pressure tests, diagnose emissions and drivability problems resulting from failure of the secondary air injection and catalytic converter systems, and properly inspect and test catalytic converter systems.

Tools and Materials

Rubber mallet
Pyrometer
Pressure gauge
Propane enrichment tool
Exhaust gas analyzer
A vehicle
Hoist

Protective Clothing

Goggles or safety glasses with side shields

Describe the vehicle being worked on:

Year ________ Make ____________________ Model ____________________

VIN ____________________ Engine type and size ____________________

Describe general operating condition:

____

____

PROCEDURE

1. Securely raise the vehicle on a hoist. Task Completed ☐

   Make sure you have easy access to the catalytic converter and that it is not HOT.

2. Using a rubber mallet, smack the exhaust pipe by the converter. Did it rattle? ☐ Yes ☐ No

   If it did, it needs to be replaced and there is no need to do any more testing. A rattle indicates loose substrate, which will soon rattle into small pieces.

3. If the converter passed this test, it doesn't mean it is in good shape. It should be checked for plugging or restrictions. Task Completed ☐

4. Lower the vehicle and open the hood. Task Completed ☐

5. Remove the O_2 sensor. Task Completed ☐

6. Install a pressure gauge into the sensor's bore. On some engines, it is not easy to do. On engines with a PFE, you can use its port to install your gauge. On other engines, you may need to fabricate a tester from an old O_2 sensor or air check valve. Task Completed ☐

7. After the gauge is in place, start the engine and hold the engine's speed at 2,000 rpm. Record the reading on the pressure gauge. _______ psi.

   You are looking for exhaust pressure under 1.25 psi. A very bad restriction would give you over 2.75 psi.

8. What does your reading tell you?

   ____

   ____

   Newer cars should have pressures well under 1.25. Some older ones can be as high as 1.75 and still be good. You will notice that if you quickly rev up the engine, the pressure goes up. This is normal. Remember, do this test at 2,000 rpm, not with the throttle wide open.

9. Remove the pressure gauge, turn off the engine, allow the exhaust to cool, and then install the O_2 sensor. Task Completed ☐

10. If the converter passed this test, you can now check its efficiency. There are three ways to do this. The first way is the delta temperature test. Start the engine and allow it to warm up. With the engine running, carefully raise the vehicle using the hoist or lift. Task Completed ☐

11. With a pyrometer, measure and record the inlet temperature of the converter. The reading was __________.

12. Now measure the temperature of the converter's outlet. The reading was __________.

13. What was the percentage increase of the temperature at the outlet compared to the temperature of the inlet? ___________

14. There should be a temperature increase of about 8% or 100 degrees at the converter's outlet. If the temperature doesn't increase by 8%, replace the converter. What are your conclusions based on this test?

____

____

15. How does temperature show the efficiency of a catalytic converter?

____

____

16. Now you can do the O_2 storage test. This is based on the fact that a good converter stores oxygen. The following test is for closed-loop feedback systems. Non-feedback systems require a different procedure. Begin by disabling the air injection system. Task Completed ☐

17. Turn on your gas analyzer and allow it to warm up. Start the engine and warm the car up as well. Task Completed ☐

18. When everything is ready, hold the engine at 2,000 rpm. Watch the exhaust readings. Record the readings:

    ______HC ______CO ______CO_2 ______O_2 ______NOx

    If the converter was cold, the readings should continue to drop until the converter reaches light-off temperature.

19. When the numbers stop dropping, check the oxygen levels. Check and record the oxygen level; the reading was __________. O_2 should be about .5 to 1%. This shows the converter is using most of the available oxygen. There is one exception to this: If there is no CO left, there can be more oxygen in the exhaust. However, it still should be less than 2.5%. It is important that you get your O_2 reading as soon as the CO begins to drop. Otherwise, good converters will fail this test. The O_2 will go way over 1.25 after the CO starts to drop.

20. If there is too much oxygen left, and no CO in the exhaust, stop the test and make sure the system has control of the air–fuel mixture. If the system is in control, use your propane enrichment tool to bring the CO level up to about .5%. Now the O_2 level should drop to zero. Task Completed ☐

21. Once you have a solid oxygen reading, snap the throttle open, and then let it drop back to idle. Check the oxygen. The reading was __________. It should not rise above 1.2%.

22. If the converter passes these tests, it is working properly. If the converter fails the tests, chances are that it is working poorly or not at all. The final converter test uses a principle that checks the converter as it is doing its actual job, converting CO and HC into CO_2 and water.

23. Allow the converter to warm up by running the engine. Task Completed ☐

24. Calibrate the gas analyzer and insert its probe into the exhaust pipe. If the vehicle has dual exhaust with a cross over, plug the side that the probe isn't in. If the vehicle has a true dual exhaust system, check both sides separately. Task Completed ☐

25. Turn off the engine and disable the ignition. Task Completed ☐

26. Crank the engine for 9 seconds as you pump the throttle. Look at the gas analyzer and record the CO_2 reading: The CO_2 for injected cars should be over 11%. If you are cranking the engine and the HC goes above 1,500 ppm, stop cranking: the converter is not working. Also stop cranking once you hit your 10 or 11% CO_2 mark; the converter is good. If the converter is bad, you should see high HC and low CO_2 at the tailpipe. What are your conclusions from this test?

___

___

___

27. Do not repeat this test more than ONE time without running the engine in between. Task Completed ☐

28. Reconnect the ignition and start the engine. Do this as quickly as possible to cool off the converter. Task Completed ☐

Problems Encountered

___

___

___

Instructor's Comments

___

___

___

ENGINE PERFORMANCE JOB SHEET 40

Checking and Refilling Diesel Exhaust Fluid

Name ______________________________ Station __________________ Date __________________

NATEF Correlation

This Job Sheet addresses the following **MLR** task:

C.5. Check and refill diesel exhaust fluid (DEF).

This Job Sheet addresses the following **AST** task:

D.11. Check and refill diesel exhaust fluid (DEF).

This Job Sheet addresses the following **MAST** task:

D.12. Check and refill diesel exhaust fluid (DEF).

Objective

Upon completion of this job sheet, you will be able to check and refill the Diesel DEF fluid.

Tools and Materials

Service information
Diesel vehicle
DEF (Diesel Exhaust Fluid)

Protective Clothing

Goggles or safety glasses with side shields

Describe the vehicle being worked on:

Year __________ Make __________________________ Model ______________________________

VIN __________________________ Engine type and size __________________________

PROCEDURE

1. Describe the purpose of the Diesel Exhaust Fluid (DEF) system. Which emission does it reduce?

____

____

2. What is the capacity of the DEF tank?

____

____

3. Does the vehicle have a DEF-level sensor in the DEF tank?

____

____

4. Which component in the diesel exhaust system uses the DEF?__________

____

5. How does the ECM determine that DEF fluid is needed?

____

____

6. At what temperature will DEF freeze?

____

____

7. How does the vehicle prevent DEF from freezing?

____

____

8. List the warnings that the vehicle gives the driver about the level of DEF in the vehicle.

____

____

9. When the DEF tank is empty, what steps does the ECM take to ensure that the driver will not ignore the warning lamp?

____

____

10. What happens if the system detects DEF of inferior quality in the DEF tank?

____

____

11. How does the ECM monitor the quality of the DEF fluid?

____

____

12. When should DEF levels be checked?

____

____

13. Describe the procedure to refill the DEF tank.

____

____

14. Describe the precautions that must be followed when working around DEF.

____

____

Problems Encountered

____

____

____

Instructor's Comments

____

____

____

ENGINE PERFORMANCE JOB SHEET 41

Checking Boost Systems

Name ______________________ Station ______________ Date ______________

NATEF Correlation

This Job Sheet addresses the following **MAST** task:

D.13. Test the operation of turbocharger/supercharger systems; determine necessary action.

Objective

Upon completion of this job sheet, you will be able to test the operation of a turbocharger/supercharger system.

Tools and Materials

Soap and water mixture

Pressure gauge

Service information

Protective Clothing

Goggles or safety glasses with side shields

Describe the vehicle being worked on:

Year __________ Make ____________________ Model ____________________

VIN ____________________ Engine type and size ____________________

Describe general operating condition:

____

____

PROCEDURE

1. Check all linkages and hoses connected to the turbocharger. Record your findings.

   ____

   ____

2. Inspect the waste gate diaphragm linkage for looseness and binding, and check the hose from the waste gate diaphragm to the intake manifold for cracks, kinks, and restrictions. Record your findings.

   ____

   ____

3. Check the coolant hoses and oil line connected to the turbocharger for leaks. Record your findings.

____

____

4. When the engine is running, does the exhaust have an odor or color? Describe the exhaust.

____

____

5. Check the condition of the engine's oil. Record your findings.

____

____

6. Check all turbocharger mounting bolts for looseness. Record your findings.

____

____

7. Pay close attention to the sound of the turbocharger when it is operating. Does it make unusual noises? Record your findings.

____

____

8. Check for exhaust leaks in the turbine housing and related pipe connections. Record your findings.

____

____

9. Check the computer with a scan tool for DTCs. Record your findings.

____

____

10. Start the engine and listen to the sound the turbo system makes while you change engine speed. Record your findings.

____

____

11. Check the air cleaner and remove the ducting from the air cleaner to turbo and look for dirt buildup or damage from foreign objects. Check for loose clamps on the compressor outlet connections. Record your findings.

____

____

12. Check the engine intake system for loose bolts or leaking gaskets. Record your findings.

____

____

13. Disconnect the exhaust pipe and look for restrictions or loose material. Examine the exhaust system for cracks, loose nuts, or blown gaskets. Record your findings.

____

____

14. Rotate the turbo shaft assembly. Does it rotate freely? Are there signs of rubbing or wheel impact damage? Record your findings.

____

____

15. Visually inspect all hoses, gaskets, and tubing for proper fit, damage, and wear. Record your findings.

____

____

16. Check the low pressure, or air cleaner, side of the intake system for vacuum leaks. Record your findings.

____

____

17. On the pressure side of the system, you can check for leaks by using soapy water. After applying the soap mixture, look for bubbles to pinpoint the source of the leak. Record your findings.

____

____

18. Connect a pressure gauge to the intake manifold to check the boost pressure. Set the gauge so you can see it during a road test. Road test the vehicle at the speed specified by the vehicle's manufacturer and observe the boost pressure. Record your findings.

____

____

19. To check the waste gate, connect a hand pressure pump and a pressure gauge to the waste gate diaphragm. Position a dial indicator against the outer end of the waste gate diaphragm rod and supply the specified pressure to the waste gate diaphragm and observe the dial indicator movement. Record your findings.

____

____

20. If the waste gate rod movement is less than specified, disconnect the rod from the waste gate valve linkage, and check the linkage for binding. Record your findings.

____

____

21. What are your service recommendations?

____

____

Problems Encountered

____

____

____

Instructor's Comments

____

____

____

ENGINE PERFORMANCE JOB SHEET 42

Check the Operation of a PCV System

Name ______________________ Station ______________ Date ______________

NATEF Correlation

This Job Sheet addresses the following **MLR** task:

D.1. Inspect, test, and service positive crankcase ventilation (PCV) filter/breather cap, valve, tubes, orifices, and hoses; perform necessary action.

This Job Sheet addresses the following **AST/MAST** tasks:

E.1. Diagnose oil leaks, emissions, and drivability concerns caused by the positive crankcase ventilation (PCV) system; determine necessary action.

E.2. Inspect, test, and service positive crankcase ventilation (PCV) filter/breather cap, valve, tubes, orifices, and hoses; perform necessary action.

Objective

Upon completion of this job sheet, you will be able to properly perform the following tasks:

1. Diagnose oil leaks, emissions, and drivability problems resulting from failure of the positive crankcase ventilation (PCV) system.
2. Inspect and test positive crankcase ventilation (PCV) filter/breather cap, valve, tubes, orifices, and hoses.

Tools and Materials

Appropriate service information

Exhaust gas analyzer

Clean engine oil

Protective Clothing

Goggles or safety glasses with side shields

Describe the vehicle being worked on:

Year ________ Make ________________ Model ________________

VIN ________________ Engine type and size ________________

Describe general operating condition:

____

____

PROCEDURE

1. Inspect the entire engine and look for signs of oil leakage. If any were found, describe their location.

   ____

   ____

2. How can an oil leak affect the operation of the engine?

   ____

   ____

3. Does the engine have a PCV valve or does it use a fixed orifice?

   ____

   ____

4. Describe the location of the PCV valve.

   ____

   ____

5. Is the PCV valve heated? If so, how is it heated?

   ____

   ____

6. Connect a scan tool and check for PCV-related DTCs. Record your findings here.

   ____

   ____

7. Prepare the exhaust analyzer, and then run the engine until normal operating temperature is reached. Record the CO reading from the exhaust analyzer. ________ %
8. Remove the PCV valve from the valve or camshaft cover. Record the CO reading now. ________%
9. Explain why there was a change in CO and what service you would recommend.

   ____

   ____

10. Place your thumb over the end of the PCV valve. Record the CO reading now. ________%

11. Explain why there was a change in CO and what service you would recommend.

____

____

12. Remove the valve from its hose and check for a vacuum. Do you feel a vacuum?

____

____

13. Hold and shake the PCV valve. Record your results.

____

____

14. If the engine has a fixed orifice, inspect it. Is it free of built-up matter? If so, what can you use to clean it out?

____

____

15. State your conclusions about the PCV system on this engine.

____

____

Problems Encountered

____

____

____

Instructor's Comments

____

____

____

ENGINE PERFORMANCE JOB SHEET 43

Check the Operation of an EGR Valve

Name ______________________________ Station ________________ Date ________________

NATEF Correlation

This Job Sheet addresses the following **AST** tasks:

E.3. Diagnose emissions and drivability concerns caused by the exhaust gas recirculation (EGR) system; determine necessary action.

E.4. Inspect, test, service and replace components of the EGR system, including EGR tubing, exhaust passages, vacuum/pressure controls, filters and hoses; perform necessary action.

This Job Sheet addresses the following **MAST** tasks:

E.3. Diagnose emissions and drivability concerns caused by the exhaust gas recirculation (EGR) system; determine necessary action.

E.6. Inspect and test electrical/electronic sensors, controls, and wiring of exhaust gas recirculation (EGR) systems; perform necessary action.

E.7. Inspect, test, service and replace components of the EGR system, including EGR tubing, exhaust passages, vacuum/pressure controls, filters and hoses; perform necessary action.

Objective

Upon completion of this job sheet, you will be able to diagnose the emissions and drivability problems caused by failure of the exhaust gas recirculation (EGR) system. You will be able to inspect and test valve, valve manifold, exhaust passages, vacuum/pressure controls, filters, and hoses of exhaust gas recirculation (EGR) systems. You will also be able to inspect and test electrical/electronic sensors, controls, and the wiring of exhaust gas recirculation (EGR) systems.

Tools and Materials

Hand-operated vacuum pump

Vacuum gauge

Scan tool for assigned vehicle

Protective Clothing

Goggles or safety glasses with side shields

Describe the vehicle being worked on:

Year __________ Make ________________________ Model ________________________

VIN ________________________ Engine type and size ________________________

Describe general operating condition:

____

____

PROCEDURE

1. Describe the type of EGR system found on the vehicle.

____

____

____

2. Is the EGR valve vacuum- or electrically operated? ____________

   Name the sensors and modules used in the operation of the EGR valve.

____

____

____

____

____

3. Do a visual inspection of the EGR system. Inspect the hoses, connectors, and exhaust tubing as necessary, are they OK? ____________

4. Can the EGR valve be opened and closed by the scan tool to check its operation? ____________

5. If so, connect the scan tool and run the engine at idle. Partially open the valve. Does the EGR opening read out on the scan tool? ____________

6. Continue to open the valve a little at a time. At what opening does the idle begin to get rough? ____________

7. Can the valve be opened enough to make the engine stall? ____________

8. If the EGR can be commanded open with the scan tool, and the engine does not run rough, what could be wrong with the EGR system? Explain your reasoning.

____

____

9. Look at the EGR monitor with the scan tool. Was it completed? If not, what DTC was set or is pending?

____

____

____

10. If a DTC was set or pending, describe what the code represents.

____

____

____

____

11. Look up the conditions necessary to run the DTC monitor. List them here.

____

____

____

____

12. If you were to drive the vehicle and the monitor passed, what would you say about the repair?

____

____

13. For DPFE EGRs on Ford vehicles, insert a tee fitting into the vacuum hose and reconnect the supply line to the transducer. Connect the vacuum gauge to the tee fitting. Start the engine. Record the vacuum reading: ____________ in. Hg

14. Actuate the solenoid with the scan tool. Did the solenoid click? ☐ Yes ☐ No

15. Did the vacuum fluctuate with the cycling of the solenoid? ☐ Yes ☐ No

Problems Encountered

____

____

____

Instructor's Comments

____

____

____

ENGINE PERFORMANCE JOB SHEET 44

Air Injection System Diagnosis and Service

Name ______________________________ Station __________________ Date __________________

NATEF Correlation

This Job Sheet addresses the following **AST** task:

E.5. Inspect and test electrical/electronically operated components and circuits of air injection systems; perform necessary action.

This Job Sheet addresses the following **MAST** tasks:

E.4. Diagnose emissions and drivability concerns caused by the secondary air injection and catalytic converter systems; determine necessary action.

E.8. Inspect and test electrical/electronically operated components and circuits of air injection systems; perform necessary action.

Objective

Upon completion of this job sheet, you will be able to inspect and test the mechanical components of secondary air injection systems, as well as be able to inspect and test the electrical/electronically operated components and circuits of air injection systems.

Tools and Materials

Vehicle with an AIR system

Exhaust gas analyzer

Vacuum gauge

Scan tool

Lab scope

Protective Clothing

Goggles or safety glasses with side shields

Describe the vehicle being worked on:

Year __________ Make __________________________ Model __________________________

VIN __________________________ Engine type and size __________________________

Describe general operating condition:

____

____

PROCEDURE

Secondary Air Injection System Diagnosis

1. Check all vacuum and delivery hoses and electrical connections in the system. Summarize the results of this check.

____

____

____

____

2. Where is the air pump located?

____

3. Describe the operation of the air pump on your assigned vehicle.

   A. When does the air pump operate?

   ____

   B. What is the purpose of the air pump?

   ____

4. How many valves are used in the air system to direct the flow of air? Describe them here.

____

____

5. What would be the consequences of the air valve diverting all the time?

____

6. What are the consequences of the air valve running at all times?

____

____

7. Has the monitor for the secondary air injection system run? If not, what code is stored or pending?

____

8. Explain when the secondary air system monitor is run by the computer.

____

____

9. Explain how the ECM knows that the air system is functional.

____

____

____

____

____

10. Check the hoses in the air system for evidence of burning. This would indicate a leak and could be the cause of excessive noise. Summarize the results of this check.

____

____

____

____

Check Valve Testing

1. All of the types of air injection systems have a one-way check valve. The valve opens to let air in but closes to keep exhaust from leaking out. If exhaust has been leaking from the valve, then an exhaust leak would exist, or the hose to the valve would be burned. Summarize the results of this check.

____

____

System Efficiency Test

1. Run the engine at idle and command the secondary air system on (enabled). Using a scan tool, measure the STFT (short term fuel trim). What were the readings?

____

____

2. Turn off the secondary air system and continue to allow the engine to idle. Again, measure and record the STFT. What were the readings?

____

Problems Encountered

Instructor's Comments

ENGINE PERFORMANCE JOB SHEET 45

Evaporative Emission Control System Diagnosis

Name ______________________ Station ______________ Date ______________

NATEF Correlation

This Job Sheet addresses the following **AST** tasks:

E.7. Inspect and test components and hoses of evaporative emissions control system; perform necessary action.

E.8. Interpret diagnostic trouble codes (DTCs) and scan tool data related to the emission control systems; determine necessary action.

This Job Sheet addresses the following **MAST** tasks:

E.5. Diagnose emissions and drivability concerns caused by the evaporative emissions control system; determine necessary action.

E.10. Inspect and test components and hoses of evaporative emissions control system; perform necessary action.

E.11. Interpret diagnostic trouble codes (DTCs) and scan tool data related to the emission control systems; determine necessary action.

Objective

Upon completion of this job sheet, you will be able to diagnose emissions and drivability problems resulting from the failure of the evaporative emissions control system, and you will also be able to inspect and test the components and the hoses of the evaporative emissions control system.

Tools and Materials

Service information
Digital multimeter
Scan tool
EVAP tester
EVAP leak tester
Hand-operated vacuum pump
Fuel cap tester

Protective Clothing

Goggles or safety glasses with side shields

Describe the vehicle being worked on:

Year ________ Make ____________________ Model ____________________

VIN ____________________ Engine type and size ____________________

PROCEDURE

NOTE: *EVAP system diagnosis varies depending on the vehicle make and model year. Always follow the service and diagnostic procedure in the vehicle manufacturer's service information.*

1. Use a scan tool and check for any EVAP-related DTCs. If an EVAP DTC is set, always correct the cause of this code before further EVAP system diagnosis. Summarize the results of your check.

   ____

   ____

2. Describe the EVAP system found on this vehicle.

   ____

   ____

3. How does this system check for leaks?

   ____

   ____

4. Check and describe the status of the EVAP monitor.

   ____

   ____

5. How many trips are necessary to run the monitor?

   ____

   ____

6. Observe the STFT. How does this relate to vapor purge?

   ____

   ____

7. When the engine is idling, what is the status of the purge solenoid?

   ____

   ____

8. Keep the scan tool connected and take the vehicle for a road test that includes all required enable criteria. LET SOMEONE ELSE DRIVE while you monitor the system from the passenger seat. What conditions does this include?

   ____

   ____

9. If the purge solenoid does not turn on during the road test, what is indicated?

   ____

   ____

10. After the road test, check all EVAP hoses for leaks, restrictions, and loose connections. Describe your findings.

____

____

11. Check the canister for cracks or damage. Describe what you found.

____

____

12. The electrical connections in the EVAP system should be checked for looseness, corroded terminals, and worn insulation. Summarize the results of your check.

____

____

13. Check for vacuum leaks in the EVAP system. A vacuum leak in any of the evaporative emission components or hoses can cause starting and performance problems, as can any engine vacuum leak. It can also elicit complaints of fuel odor. Summarize the results of your check.

____

____

14. Close the purge and intake ports at the canister and apply low air pressure (about 2.8 psi [19.6 kPa]) to the vent port. Was the canister able to hold that pressure for at least one minute? What does this indicate?

____

____

15. Measure the voltage drop across the canister purge solenoid and compare your readings to specifications. If the readings are outside specifications, replace the charcoal canister assembly.

____

____

16. The canister purge solenoid winding may be checked with an ohmmeter. Find the specifications in the service information and compare your measurements to them. Summarize the results of your check.

____

____

17. Remove the tank pressure control valve. Try to blow air through the valve with your mouth from the tank side of the valve. What happened and what does this indicate?

____

____

18. Connect a vacuum pump to the vacuum fitting on the pressure control valve and apply 10 in. Hg to the valve. Now try to blow air through the valve from the tank side. What happened and what does this indicate?

____

____

19. Allow the engine to run until it reaches normal operating temperature. Connect the purge flow tester's flow in series with the engine and evaporative canister. Zero the gauge of the tester with the engine off, and then start the engine. With the engine at idle, turn on the tester and record the purge flow rate and accumulated purge volume. What did you measure?

____

____

20. Gradually increase engine speed to about 2500 rpm and record the purge flow. What did you measure?

____

____

21. What can you conclude from the two previous tests?

____

____

22. How much fuel is in the vehicle?

____

____

23. Use the pressure chart that accompanies the EVAP leak tester and determine how much pressure the tester should be set at. Record this amount.

____

____

24. Connect the tester to the EVAP service port. Where is this located?

____

____

25. Set the scan tool for an EVAP test. Adjust the output pressure from the tester and observe how much pressure the system can hold. What does this indicate?

____

____

26. What method will you use to identify the location of a leak? Briefly describe the procedure for doing this.

____

____

27. What equipment will you use to check the fuel cap? Briefly describe the procedure for doing this.

____

____

28. Remove the fuel cap and inspect the filler neck. Describe its condition.

____

____

29. If the fuel tank has a pressure and vacuum valve in the filler cap, check these valves for dirt contamination and damage. The cap may be washed in clean solvent. When the valves are sticking or damaged, replace the cap. Summarize the results of your check.

____

____

30. What are your conclusions about this EVAP system?

____

____

Problems Encountered

____

____

____

Instructor's Comments

____

____

____

— NOTES —